Solution Manual for

Quantum Mechanics

2nd Edition

Solution Manual for
Quantum Mechanics

2nd Edition

Ahmed Ishtiaq
Fayyazuddin
Riazuddin

National Centre for Physics, Pakistan

World Scientific

NEW JERSEY · LONDON · SINGAPORE · BEIJING · SHANGHAI · HONG KONG · TAIPEI · CHENNAI

Published by

World Scientific Publishing Co. Pte. Ltd.

5 Toh Tuck Link, Singapore 596224

USA office: 27 Warren Street, Suite 401-402, Hackensack, NJ 07601

UK office: 57 Shelton Street, Covent Garden, London WC2H 9HE

Library of Congress Cataloging-in-Publication Data
Ishtiaq, Ahmed, author.
  Solution manual for Quantum mechanics (2nd edition) / Ahmed Ishtiaq (National Centre for
Physics, Pakistan), Fayyazuddin (National Centre for Physics, Pakistan), Riazuddin (National Centre
for Physics, Pakistan).
    pages cm
  Includes bibliographical references and index.
  ISBN 978-9814541886 (softcover : alk. paper)
  1. Quantum theory--Problems, exercises, etc. I. Fayyazuddin, 1930– author.
Quantum mechanics. II. Riazuddin. author. Quantum mechanics. III. Title. IV. Title: Quantum
mechanics (2nd edition).
  QC174.15.I84 2014
  530.12--dc23

                                            2013046034

British Library Cataloguing-in-Publication Data
A catalogue record for this book is available from the British Library.

Typeset by Stallion Press
Email: enquiries@stallionpress.com

Printed in Singapore

Contents

Chapter 1

Breakdown of Classical Concepts

The following formulae are relevant for the solution of the problems in this chapter.

For photons

$$\nu\lambda = c$$

$$E = h\nu = 2\pi\hbar\nu = \hbar\omega; \quad \omega = 2\pi\nu \qquad (1.1)$$

$$p = \frac{E}{c} = h\frac{\nu}{c} = \frac{h}{\lambda} = \frac{2\pi\hbar}{\lambda} = \hbar k; \quad k = \frac{2\pi}{\lambda}$$

For particles

$$E = h\nu = \hbar\omega$$

$$\nu\lambda = v$$

$$p = \frac{h}{\lambda} = \frac{2\pi\hbar}{\lambda} = \hbar k$$

$$\text{group velocity} = \frac{d\omega}{dk}$$

$\hbar = 6.582\times10^{-22}$ MeV.s, $c\hbar = 1.97\times10^{-13}$ MeV.m, $c \approx 3\times10^8$ m.s^{-1}

Q1.1 Find the energy and momentum of an X-ray photon whose frequency is 5×10^{18} cycles sec^{-1}.

Solution:

$$E = h\nu = 2\pi\hbar(5 \times 10^{18}\,\text{s}^{-1}) = 20.68 \times 10^{-3}\,\text{MeV}$$

$$p = \frac{E}{c} = 20.68 \times 10^{-3}\,\text{MeV}/c \qquad (1.2)$$

1

Q1.2 (a) The velocity of ripples on a liquid surface is $\sqrt{2\pi S/\lambda\rho}$, where S is the surface tension and ρ the density of the liquid. Find the group velocity of these waves.

(b) Velocity of ocean waves is $\sqrt{g\lambda/2\pi}$, where g is the acceleration due to gravity. Find the group velocity of ocean waves.

Solution: (a)

$$v = \sqrt{\frac{2\pi S}{\lambda\rho}}$$

and

$$v = \nu\lambda$$

Thus

$$\nu = \sqrt{\frac{2\pi S}{\lambda^3\rho}}$$

Now

$$v_g = \frac{d\omega}{dk} = \frac{d\nu}{d(1/\lambda)} = -\lambda^2\frac{d\nu}{d\lambda}$$

$$= -\lambda^2\sqrt{\frac{2\pi S}{\rho}}(-3/2)(\lambda)^{-5/2} = \frac{3}{2}\sqrt{\frac{2\pi S}{\rho\lambda}} = \frac{3}{2}v$$

(b) For the ocean wave

$$v = \sqrt{\frac{g\lambda}{2\pi}}$$

we get

$$v_g = \frac{1}{2}v$$

Q1.3 Given that $A(k) = \left(\frac{1}{2K}\right)^{\frac{1}{2}}$, $|k| \leq K$, $A(k) = 0$, $|k| > K$. Find $\psi(x)$. Make a sketch of $A(k)$ and $\psi(x)$ and hence show that $\Delta k \Delta x > 1$.

Solution: To find $\psi(x)$ use Eq. (1.1b) from the book,

$$\psi(x) = \int_{-K}^{K} \frac{1}{\sqrt{2K}} e^{ikx} dk$$

$$= \sqrt{2K} \left(\frac{\sin Kx}{Kx} \right) \tag{1.3}$$

A sketch of $A(k)$ and $\psi(x)$ are shown in Figs. 1.1 and 1.2 respectively. It is clear from these figures that $\Delta k = 2K$

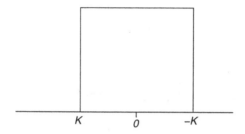

Fig. 1.1   Sketch of $A(k)$.

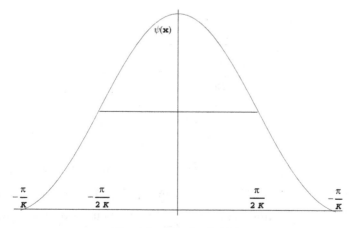

Fig. 1.2   Sketch of $\psi(x)$.

and $\Delta x = \frac{\pi}{K}$, so that $\Delta k \Delta x = 2\pi > 1$, in conformity with the well known relation $\Delta x \Delta k \geq 1$ of wave optics.

Q1.4 For a Guassian wave packet

$$\psi(x,t) = Ne^{-x^2/2\delta^2} e^{i(k_0 x - \omega_0 t)}$$

Show that

$$A(k) = \frac{N}{\sqrt{2\pi}} \delta e^{-\delta^2 (k-k_0)^2/2}$$

Make a sketch of $|\psi(x)|^2 = |\psi(x,0)|^2$ and $|A(k)|^2$ ·and hence show that

$$\Delta x \Delta k = 1/2 \qquad (1.4)$$

Solution:

$$\psi(x,t) = Ne^{-\frac{x^2}{2\delta^2}} e^{i(k_0 x - \omega t)}$$

$$\psi(x,0) = Ne^{-\frac{x^2}{2\delta^2}} e^{ik_0 x}$$

$$A(k) = \frac{1}{2\pi} \int_{-\infty}^{\infty} e^{-ikx} \psi(x,0) dx$$

$$= \frac{1}{2\pi} N \int_{-\infty}^{\infty} e^{-\frac{x^2}{2\delta^2}} e^{-i(k-k_0)} dx$$

$$= \frac{N}{2\pi} e^{-\frac{\delta^2}{2}(k-k_0)^2} \int_{-\infty}^{\infty} e^{-\frac{Y^2}{2\delta^2}} dY$$

$$= \frac{N}{\sqrt{2\pi}} \delta e^{-\frac{\delta^2}{2}(k-k_0)^2}$$

where we have put $x + i\delta^2(k - k_0) = Y$ and Gaussian integral $\int_{-\infty}^{\infty} e^{-\frac{Y^2}{2\delta^2}} dY = \sqrt{2\pi}\delta$. From Fig. 1.3 and Fig. 1.4, $\Delta x = \frac{\delta}{\sqrt{2}}$ and $\Delta k = \frac{1}{\sqrt{2}\delta}$ and their product gives $\Delta x \Delta k = 1/2$.

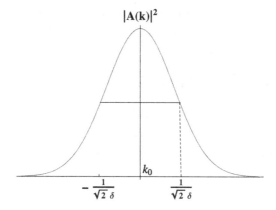

Fig. 1.3   Sketch of $A(k)$.

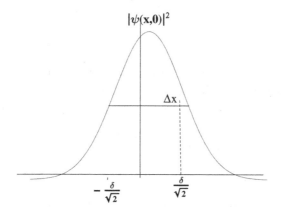

Fig. 1.4   Sketch of $\psi(x)$.

Q1.5 Light of wave length $\lambda = 10$ meters falls on the surface of a metal. If the work function of metal is 3.64 eV, is an electron emitted or not?

Solution:

$$E = h\nu = h\frac{c}{\lambda} = \frac{2\pi c\hbar}{\lambda} = \frac{2\pi(1.97 \times 10^{-13})\,\text{MeV.m}}{10\,\text{m}}$$

$$= 1.24 \times 10^{-12}\,\text{MeV} = 1.24 \times 10^{-6}\,\text{eV} < W$$

Hence no electron is emitted.

Q1.6 Find the velocity of a proton whose de Broglie wave length is equal to that of a 1 keV X-ray.

Solution: For X-ray photon

$$E = h\nu = \frac{2\pi\hbar c}{\lambda}$$

$$\lambda = \frac{2\pi\hbar c}{E}$$

For the proton, the de Broglie wave length

$$\lambda = \frac{h}{p} = \frac{2\pi\hbar}{m_p v_p}$$

Thus

$$v_p = \frac{2\pi\hbar}{m_p \lambda} = \frac{cE}{m_p c^2} = \frac{3 \times 10^8 \times 10^{-3} \,\text{MeV.m.s}^{-1}}{938 \,\text{MeV}}$$

$$= 3.2 \times 10^2 \,\text{m.s}^{-1}$$

Q1.7 A photon whose initial frequency was $1.5 \times 10^{19} \,\text{sec}^{-1}$ emerges from a collision with an electron with a frequency $1.2 \times 10^{19} \,\text{sec}^{-1}$. How much kinetic energy was imparted to electron?

Solution: From energy conservation

$$h\nu_i = h\nu_f + K$$

where K is the recoil energy of the electron.
Thus

$$K = h(\nu_i - \nu_f) = h(1.5 - 1.2)10^{19} \,s^{-1}$$

$$= 2\pi(6.582 \times 10^{-22} \,\text{MeV.s})(0.3 \times 10^{19})\,s^{-1} = 12.4 \,\text{keV}$$

Q1.8 Derive a formula expressing de Broglie wave length (in Å) of an electron in terms of potential difference V (in volts) through which it is accelerated.

Solution: The de Broglie relation:

$$\lambda = \frac{h}{m_e v} = \frac{2\pi\hbar}{m_e v}$$

Now $m_e c^2 = 0.5 \,\text{MeV}$.

Hence

$$\lambda = \frac{2\pi\hbar c}{10^3 \, \text{eV}} = \frac{2\pi(1.97 \times 10^{-13}) \, \text{MeV.m}}{10^{-3} \, \text{MeV}}$$

$$= 12.37 \times 10^{-10} \, \text{m} = 12.37 \, \text{Å}$$

Q1.9 Show that velocity v of the electron in the first Bohr orbit of hydrogen atom is given by

$$v/c = e^2/\hbar c = 1/137.$$

Solution: For the lowest level, the Bohr quantization rule of the angular momentum:

$$L = ma_0 v = \hbar$$

$$v = \frac{\hbar}{ma_0} \tag{1.5}$$

where a_0 is called Bohr's radius

$$a_0 = \frac{\hbar^2}{me^2} \tag{1.6}$$

Using the value of a_0 from Eq. (1.6) in Eq. (1.5) and multiplying both sides with $1/c$,

$$\frac{v}{c} = \frac{e^2}{\hbar c}$$

Q1.10 A beam of $100 \, \text{MeV}$ electrons travels a distance of 10 meters. If the width Δx of initial packet is $10^{-2} \, \text{cm}$, calculate the spread in the wave packet in traveling this distance and show that this spread is much less than Δx.

Solution: The spread of the wave packet (c.f. Eq. (1.20) of the text) is given by

$$t\left(\frac{d^2\omega}{dk^2}\right)\Delta k \approx t\left(\frac{d^2\omega}{dk^2}\right)\frac{1}{\Delta x}$$

Now a $100 \, \text{MeV}$ electron is a relativistic particle

Thus

$$E^2 = c^2 p^2 + m_e^2 c^4$$

$$\hbar^2 \omega^2 = c^2 \hbar^2 k^2 + m_e^2 c^4$$

$$\therefore \quad 2\hbar^2 \omega d\omega = 2c^2 \hbar^2 k dk$$

Thus

$$\frac{d^2\omega}{dk^2} = c^2 \left(\frac{1}{\omega} - \frac{k}{\omega^2} \frac{d\omega}{dk} \right)$$

$$= \frac{c^2 \hbar}{E^3} m_e^2 c^4 = \frac{c\hbar}{E} \left(\frac{m_e c^2}{E} \right)^2 c$$

Hence

$$t \left(\frac{d^2\omega}{dk^2} \right) \frac{1}{\Delta x} = \frac{lc}{v} \left(\frac{c\hbar}{E} \right) \left(\frac{m_e c^2}{E} \right)^2 \frac{1}{\Delta x}$$

Now

$$E = 100.5 \,\text{MeV}, \quad m_e c^2 = 0.5 \,\text{MeV},$$

$$v \approx c, \quad l = 10 \,\text{m}, \quad \Delta x = 10^{-4} \,\text{m}$$

Thus

$$t \left(\frac{d^2\omega}{dk^2} \right) \frac{1}{\Delta x} \ll \Delta x.$$

Q1.11 Find the de Broglie wave length of an electron accelerated through a 10 Volts potential difference. What is its velocity?

Solution:

$$\lambda = \frac{h}{m_e v} = \frac{2\pi \hbar}{m_e v}$$

$$\frac{1}{2} m_e v^2 = 10 \,\text{eV}$$

$$v = \sqrt{\frac{20}{m_e} \text{eV}}$$

$$\lambda = \frac{2\pi\hbar c}{\sqrt{20 m_e c^2 \text{eV}}} = \frac{2\pi\hbar c}{\sqrt{10\,\text{MeV} \times \text{eV}}}$$

$$= \frac{2\pi\hbar c}{\sqrt{10 \times 10^{-3}\,\text{MeV}}} = \frac{(2\pi)(1.97) \times 10^{-10}}{\sqrt{10}}\,\text{m}$$

$$= 3.91 \times 10^{-10}\,\text{m} = 3.91\,\text{Å}$$

Q1.12 A light of wave length $2500\,\text{Å}$ falls on a surface of metal whose work function is $3.64\,\text{eV}$. What is the kinetic energy of the electrons emitted from the surface? If the intensity of light is $4.0\,\text{W/m}^2$, find the average number of photons per unit time per unit area that strike the surface.

Solution: Energy of the photon of wave length $2500\,\text{Å}$:

$$E = h\nu = 2\pi\hbar\frac{c}{\lambda} = 4.95\,\text{eV}$$

$$E = W + T$$

$$T = E - W = 1.31\,\text{eV}$$

$$n h\nu = I$$

$$I = 4\frac{W}{\text{m}^2} = 4J \times 10^{-4}\,\text{cm}^{-2}$$

Hence

$$n = \frac{4 \times 10^{-4}\,\text{J}}{4.95\,\text{eV}}\,\text{cm}^{-2} = \frac{4 \times 10^{-4}\,\text{J}}{4.95 \times 1.602 \times 10^{-19}\,\text{J}}\,\text{cm}^{-2}$$

$$= 5 \times 10^{14}\,\text{photons per cm}^2 \tag{1.7}$$

Chapter 2

Quantum Mechanical Concepts

Q2.1 If the excited level in an atom lasts for about 10^{-10} sec., what is the order of magnitude of electron energy spread measured in eV?

Solution: From the energy–time uncertainty relation

$$\Delta E \sim \frac{\hbar}{\Delta t} = 6.6 \times 10^{-6}\,\text{eV}$$

Q2.2 What is the approximate momentum imparted to a proton initially at rest by a measurement which locates its position within 10^{-11} meters?

Solution: From the position momentum uncertainty relation

$$\Delta p \sim \frac{\hbar}{\Delta x} = 1.96 \times 10^{-2}\,\text{MeV}/c$$

Q2.3 A particle has an uncertainty in its position $\Delta x = 2a$. The magnitude of the momentum must be at least as large as the uncertainty in the momentum. Hence find an estimate of its energy.

Solution: The energy of the particle is given by

$$E = \frac{p^2}{2m} \qquad\qquad (2.1)$$

Now

$$\Delta p \geq \frac{\hbar}{\Delta x} \qquad\qquad (2.2)$$

Using $\Delta x = 2a$, $p \sim \Delta p$, we get from Eqs. (2.2) and (2.1):

$$E = \frac{\hbar^2}{8ma^2} \qquad (2.3)$$

Q2.4 Show that

$$u(x) = e^{-x^2/2} \quad \text{and} \quad u(x) = 2xe^{-\frac{1}{2}x^2}$$

are eigenfunctions of the operator

$$\hat{A}\left(x, \frac{\partial}{\partial x}\right) \equiv -\frac{d^2}{dx^2} + x^2.$$

Find the corresponding eigenvalues.

Solution: For the wave function $u(x) = e^{-x^2/2}$:

$$\hat{A}u(x) = \left(-\frac{d^2}{dx^2} + x^2\right)e^{-x^2/2} = e^{-x^2/2} = u(x)$$

Thus $u(x) = e^{-x^2/2}$ is an eigenfunction of $\hat{A}$ with eigenvalue 1. Similarly for $u(x) = 2xe^{-x^2/2}$:

$$\hat{A}u(x) = \left(-\frac{d^2}{dx^2} + x^2\right)(2xe^{-x^2/2})$$

$$= 3(2xe^{-x^2/2}) = 3u(x)$$

Thus $u(x) = 2xe^{-x^2/2}$ is an eigenfunction of $\hat{A}$ with eigenvalue 3.

Q2.5 Show that for a wave function represented by a plane wave

$$\psi(\mathbf{r}) = e^{i\mathbf{p}\cdot\mathbf{r}/\hbar} \qquad (2.4)$$

the probability current density $\mathbf{S}$ is given by

$$\mathbf{S}(\mathbf{r}, t) = \frac{\mathbf{p}}{m} = \mathbf{v},$$

where m is the mass of the particle.

Solution: The probability current density can be calculated by the following relation

$$\mathbf{S} = \frac{\hbar}{2mi}[\psi^*\boldsymbol{\nabla}\psi - \psi\boldsymbol{\nabla}\psi^*] \qquad (2.5)$$

The wave function for plane wave is

$$\psi(\vec{r}) = e^{i\mathbf{p}\cdot\mathbf{r}/\hbar} \tag{2.6}$$

Eq. (2.5) then gives

$$S_x = \frac{\hbar}{2mi}\left[\psi^*\frac{\partial}{\partial x}\psi - \psi\frac{\partial}{\partial x}\psi^*\right]$$

$$= \frac{\hbar}{2mi}\left[\frac{i}{\hbar}p_x - \left(-\frac{i}{\hbar}\right)p_x\right]$$

$$= \frac{p_x}{m}$$

$$= v_x$$

Similarly $S_y = p_y/m = v_y$ and $S_z = p_z/m = v_z$
Thus

$$\mathbf{S} = S_x\mathbf{i} + S_y\mathbf{j} + S_z\mathbf{k}$$

$$= \frac{\mathbf{p}}{m} \tag{2.7}$$

$$= \mathbf{v}$$

Q2.6 In β-decay, electrons are emitted from the nuclei with energy of few MeV. Take the energy 1 MeV and size of the nucleus 10^{-13} cm. Use the uncertainty principle to show that electrons cannot be contained in the nucleus before the decay.

Solution: To find the momentum use the relativistic energy–momentum relation

$$E^2 = c^2p^2 + m^2c^4 \tag{2.8}$$

Using $E = 1$ MeV in Eq. (2.8), one gets $p = 0.87$ MeV/c. Using the value of p in the uncertainty relation

$$\Delta x \geq \frac{\hbar c}{pc}$$

gives $\Delta x \geq 2.26 \times 10^{-11}$ cm, which is larger than the size of the nucleus (10^{-13} cm).
Thus the electrons cannot be contained in the nucleus.

Chapter 3

Basic Postulates of Quantum Mechanics

Q3.1 If $\langle \widehat{A} \rangle$ denotes the average value for a large number of measurements of an operator $\widehat{A}$ for an arbitrary state function ψ, show that $\langle A \rangle$ is real if $\widehat{A}$ is hermitian.

Solution: For a hermitian operator (c.f. Eq. 3.9):

$$(\psi|\hat{A}|\phi) = (\phi|\hat{A}^\dagger \psi)^*$$
$$= (\phi|A\psi)^*$$

Replacing ϕ by ψ in the above equation, one gets for the average value of hermitian operator

$$\langle \hat{A} \rangle = (\psi|\hat{A}\psi) = (\psi|A\psi)^*$$

Q3.2 The state function for a free particle moving in x-direction is given by

$$\psi(x) = Ne^{(-x^2/2\delta^2 + ip_0 x/\hbar)}.$$

Normalise this wave function. Find the state function $\phi(p)$ in momentum space.

(i) Show that for the state $\psi(x)$ given above

$$\langle x \rangle = 0$$
$$\langle p \rangle = p_0$$

$$\langle x^2 \rangle = \frac{\delta^2}{2}$$

$$\langle p^2 \rangle = p_0^2 + \frac{1}{2}\frac{\hbar^2}{\delta^2}.$$

Hence show that

$$(\overline{\Delta x})^2 = \langle (x - \langle x \rangle)^2 \rangle$$

$$= \frac{\delta^2}{2}$$

$$(\overline{\Delta p})^2 = \langle (p - \langle p \rangle)^2 \rangle$$

$$= \frac{1}{2}\frac{\hbar^2}{\delta^2}$$

so that

$$\Delta x \Delta p = \frac{1}{2}\hbar.$$

(ii) Using the relations

$$\langle p \rangle = \int p |\phi(p)|^2 dp$$

$$\langle p^2 \rangle = \int p^2 |\phi(p)|^2 dp$$

show that

$$\langle p \rangle = p_0$$

$$\langle p^2 \rangle = p_0^2 + \frac{1}{2}\frac{\hbar^2}{\delta^2}.$$

$$\left(\text{Useful integral } \int_{-\infty}^{+\infty} x^{2n} e^{-\alpha x^2} dx = \frac{1}{\alpha^{n+\frac{1}{2}}} \frac{\sqrt{\pi}(2n)!}{2^{2n} n!} \right)$$

Solution: The normalization condition is

$$\int \psi(x)^* \psi(x) dx = 1 \rightarrow \int |\psi(x)|^2 dx = 1$$

gives

$$|N|^2 \int e^{-\frac{x^2}{\delta^2}} dx = 1 \qquad (3.1)$$

Using the integral given in the problem for $n = 0$, $\alpha = \frac{1}{\delta^2}$

$$|N|^2 \sqrt{\frac{\pi}{1/\delta^2}} = 1$$

Thus

$$N = \left(\frac{1}{\pi\delta^2}\right)^{1/4}$$

where we have selected the phase so that N is positive number. Hence the normalized wave function is

$$\psi(x) = \left(\frac{1}{\pi\delta^2}\right)^{1/4} e^{\left(-\frac{x^2}{2\delta^2} + \frac{ip_0 x}{\hbar}\right)}$$

Find $\langle x \rangle$:

$$\langle x \rangle = \int \psi^*(x)\hat{x}\psi(x)dx,$$

$$= \left(\frac{1}{\pi\delta^2}\right)^{1/2} \int_{-\infty}^{+\infty} x e^{-\frac{x^2}{\delta^2}} dx$$

$$= 0$$

since the integral vanishes, as the integrand is odd function of x.

Find $\langle p \rangle$:

$$\langle p \rangle = \int_{-\infty}^{+\infty} \psi^*(x)\hat{p}\psi(x)dx,$$

$$= \left(\frac{1}{\pi\delta^2}\right)^{1/2} \int_{-\infty}^{+\infty} e^{\left(-\frac{x^2}{2\delta^2} - \frac{ip_0 x}{\hbar}\right)}$$

$$\times \left(-i\hbar\frac{\partial}{\partial x}\right) e^{\left(-\frac{x^2}{2\delta^2} + \frac{ip_0 x}{\hbar}\right)} dx$$

$$= \left(\frac{1}{\pi\delta^2}\right)^{1/2} \int_{-\infty}^{+\infty} e^{-\frac{x^2}{\delta^2}} (-i\hbar) \left(\frac{ip_0}{\hbar} - \frac{x}{\delta^2}\right) dx \quad (3.2)$$

$$= \left(\frac{1}{\pi\delta^2}\right)^{1/2} p_0 (\pi\delta^2)^{1/2} = p_0$$

Since the second term in the integrand in Eq. (3.2) gives zero as noted above.

Find $\langle x^2 \rangle$:

$$\langle x^2 \rangle = \int_{-\infty}^{+\infty} \psi^*(x) x^2 \psi(x) dx$$

$$= \left(\frac{1}{\pi\delta^2}\right)^{1/2} \int_{-\infty}^{+\infty} x^2 e^{\left(-\frac{x^2}{\delta^2}\right)} dx$$

$$= \frac{\delta^2}{2} \quad (3.3)$$

Find $\langle p^2 \rangle$: The expectation value of p^2 is

$$\langle p^2 \rangle = \int_{-\infty}^{+\infty} \psi^*(x) \hat{p}^2 \psi(x) dx,$$

$$= \int_{-\infty}^{+\infty} e^{\left(-\frac{x^2}{2\delta^2} - \frac{ip_0 x}{\hbar}\right)} \left(-i\hbar \frac{\partial}{\partial x}\right)^2 e^{\left(-\frac{x^2}{2\delta^2} + \frac{ip_0 x}{\hbar}\right)} dx$$

$$= \frac{1}{\sqrt{\pi}\delta} (-i\hbar)^2 \int_{-\infty}^{+\infty} e^{-x^2/\delta^2}$$

$$\times \left[-\frac{1}{\delta^2} + \frac{x^2}{\delta^4} + \frac{2ip_0}{\hbar} \frac{x}{\delta^2} - \frac{p_0^2}{\hbar^2}\right] dx$$

$$= \frac{1}{\sqrt{\pi}\delta} \hbar^2 \left[\left(\frac{1}{\delta^2}\delta - \frac{1}{\delta^4}\frac{\delta^3}{2} + \frac{p_0^2}{\hbar^2}\delta\right) \sqrt{\pi}\right]$$

$$= p_0^2 + \frac{1}{2}\frac{\hbar^2}{\delta^2}$$

Now

$$(\overline{\Delta x})^2 = \langle (x - \langle x \rangle)^2 \rangle = \langle x^2 - 2x\langle x \rangle + \langle x \rangle^2 \rangle$$

$$= \langle x \rangle^2 = \frac{\delta^2}{2}$$

Similarly,

$$(\overline{\Delta p})^2 = \langle p^2 - 2p\langle p\rangle + \langle p\rangle^2\rangle$$

$$= \frac{\hbar^2}{2\delta^2}$$

Hence

$$\Delta x \Delta p = \frac{1}{2}\hbar$$

i.e. for a Gaussian wave packet $\Delta x \Delta p$ has the minimum value.

(ii) The wave function $\phi(p)$ in the momentum space is the Fourier transform of $\psi(x)$.

$$\phi(p) = \frac{1}{\sqrt{2\pi\hbar}} \int e^{-ipx/\hbar}\psi(x)dx$$

$$= \frac{1}{\sqrt{2\pi\hbar}} \left(\frac{1}{\pi\delta^2}\right)^{1/4} \int e^{-ipx/\hbar}e^{\left(-\frac{x^2}{2\delta^2}+ip_ox\right)}dx$$

$$= \frac{1}{\sqrt{2\pi\hbar}} \left(\frac{1}{\pi\delta^2}\right)^{1/4} e^{-\frac{\delta^2}{2\hbar^2}(p-p_0)^2} \int_{-\infty}^{\infty} e^{-\frac{X^2}{2\delta^2}}dX$$

where

$$X = \left(x + \frac{i\delta^2(p-p_0)}{\hbar}\right)$$

Thus

$$\phi(p) = \left(\frac{\delta}{\sqrt{\pi}\hbar}\right)^{1/2} e^{-\frac{\delta^2}{2\hbar^2}(p-p_0)^2}$$

Now

$$\langle p\rangle = \int p|\phi(p)|^2 dp$$

Using the wave function $\phi(p)$ from the previous equation in the above equation we get

$$\langle p\rangle = \frac{\delta}{\sqrt{\pi}\hbar} \int_{-\infty}^{+\infty} pe^{-\frac{\delta^2}{\hbar^2}(p-p_0)^2} dp$$

Put $P = p - p_0$, $dP = dp$, hence

$$\langle p \rangle = \frac{\delta}{\sqrt{\pi}\hbar} \int_{-\infty}^{+\infty} (P + p_0) e^{-\frac{\delta^2}{\hbar^2}P^2} dP$$

$$= \frac{\delta}{\sqrt{\pi}\hbar} p_0 \frac{\sqrt{\pi}\hbar}{\delta} = p_0$$

$$\langle p^2 \rangle = \int p^2 |\phi(p)|^2 dp$$

$$= \frac{\delta}{\sqrt{\pi}\hbar} \int (P^2 + 2p_0 P + p_0^2) e^{-P^2 \frac{\delta^2}{\hbar^2}} dP$$

$$= \frac{\delta}{\sqrt{\pi}\hbar} \left[\frac{\sqrt{\pi}}{2} \left(\frac{\hbar}{\delta} \right)^3 + p_0^2 \frac{\sqrt{\pi}\hbar}{\delta} \right]$$

$$= p_0^2 + \frac{1}{2} \frac{\hbar^2}{\delta^2} \tag{3.4}$$

Q3.3 Show that for a particle of mass m, moving in a potential $V(\mathbf{r})$,

$$[H, \mathbf{r}] = -i\hbar \frac{\hat{\mathbf{p}}}{m}$$

$$[\hat{\mathbf{p}}, H] = [\hat{\mathbf{p}}, V(\mathbf{r})] = -i\hbar \nabla V$$

Using the above results and the fact that H is hermitian, show that

$$m \frac{d}{dt} \langle \mathbf{r} \rangle = \langle \mathbf{p} \rangle$$

$$\frac{d}{dt} \langle \mathbf{p} \rangle = -\langle \nabla V \rangle.$$

(Note that Newton's law is valid for expectation values.)

Solution: For the given Hamiltonian

$$[H, \mathbf{r}] = \left[\frac{\hat{\mathbf{p}}^2}{2m} + V(\mathbf{r}), \mathbf{r} \right] = \left[\frac{\hat{\mathbf{p}}^2}{2m}, \mathbf{r} \right]$$

because $V(\mathbf{r})$ is only function of $\mathbf{r}$. Now

$$\left[\frac{\hat{\mathbf{p}}^2}{2m}, \mathbf{r} \right] = \frac{\hat{\mathbf{p}}}{2m} [\hat{\mathbf{p}}, \mathbf{r}] + [\hat{\mathbf{p}}, \mathbf{r}] \frac{\hat{\mathbf{p}}}{2m}$$

Using

$$[\widehat{p}_i, \widehat{x}_j] = -i\hbar\delta_{ij}$$

we get

$$[\widehat{\mathbf{p}}, \mathbf{r}] = -i\hbar$$

Hence

$$[H, \mathbf{r}] = -i\hbar\frac{\widehat{\mathbf{p}}}{m} \tag{3.5}$$

- Now we show

$$[\widehat{\mathbf{p}}, H] = -i\hbar\nabla V(\mathbf{r})$$

The L.H.S gives

$$\left[\widehat{\mathbf{p}}, \frac{\widehat{\mathbf{p}}^2}{2m} + V(\mathbf{r})\right] = -i\hbar[\nabla, V(\mathbf{r})]$$

where we have used $\widehat{\mathbf{p}} = -i\hbar\nabla$. Now applying this commutation relation on an arbitrary wave function $\psi(\mathbf{r})$, we get

$$[\nabla, V(\mathbf{r})]\psi(\mathbf{r}) = \nabla(V(\mathbf{r})\psi(\mathbf{r})) - V(\mathbf{r})\nabla\psi(\mathbf{r})$$
$$= (\nabla V(\mathbf{r}))\psi(\mathbf{r}) + V(\mathbf{r})\nabla\psi(\mathbf{r}) - V(\mathbf{r})\nabla\psi(\mathbf{r})$$
$$= (\nabla V(\mathbf{r}))\psi(\mathbf{r})$$

Hence

$$[\widehat{\mathbf{p}}, H] = -i\hbar\nabla V(\mathbf{r}) \tag{3.6}$$

- Now we show

$$m\frac{d}{dt}\langle\mathbf{r}\rangle = \langle\mathbf{p}\rangle$$

If we take the expectation value of Eq. (3.5), we get

$$\langle[H, \mathbf{r}]\rangle = -\frac{i\hbar}{m}\langle\widehat{\mathbf{p}}\rangle \tag{3.7}$$

where

$$\langle [H, \mathbf{r}] \rangle = \int \psi^*(\mathbf{r}, t)[H, \mathbf{r}]\psi(\mathbf{r}, t)d\mathbf{r}$$

$$= \int (H\psi(\mathbf{r}, t))^* \mathbf{r}\psi(\mathbf{r}, t)d\mathbf{r}$$

$$- \int \psi^*(\mathbf{r}, t)\mathbf{r}(H\psi(\mathbf{r}, t)d\mathbf{r})$$

where we have used $H^\dagger = H$. Now using the Schrödinger equation

$$i\hbar\frac{\partial\psi(\mathbf{r}, t)}{\partial t} = H\psi(\mathbf{r}, t) \qquad (3.8)$$

we get

$$\langle [H, \mathbf{r}] \rangle = \int \left(i\hbar\frac{\partial\psi(\mathbf{r}, t)}{\partial t} \right)^* \mathbf{r}\psi(\mathbf{r}, t)d\mathbf{r}$$

$$- \int \psi^*(\mathbf{r}, t)\mathbf{r} \left(i\hbar\frac{\partial\psi(\mathbf{r}, t)}{\partial t} \right) d\mathbf{r}$$

$$= \int \left(-i\hbar\frac{\partial\psi^*(\mathbf{r}, t)}{\partial t} \right) \mathbf{r}\psi(\mathbf{r}, t)d\mathbf{r}$$

$$- \int \psi^*(\mathbf{r}, t)\mathbf{r} \left(i\hbar\frac{\partial\psi(\mathbf{r}, t)}{\partial t} \right) d\mathbf{r}$$

Since $\mathbf{r}$ is independent of t, we can write

$$\langle [H, \mathbf{r}] \rangle = -i\hbar\frac{d}{dt}\int \psi^*(\mathbf{r}, t)\mathbf{r}\psi(\mathbf{r}, t)d\mathbf{r} = -i\hbar\frac{d}{dt}\langle \mathbf{r} \rangle$$

Hence from Eq. (3.7) we get

$$\frac{d}{dt}\langle \mathbf{r} \rangle = \frac{\langle \widehat{\mathbf{p}} \rangle}{m}$$

- Similarly, from Eq. (3.6) we can write

$$\langle [\widehat{\mathbf{p}}, H] \rangle = -i\hbar\langle \nabla V(\mathbf{r}) \rangle \qquad (3.9)$$

where

$$\langle [\widehat{\mathbf{p}}, H] \rangle = \int \psi^*(\mathbf{r}, t)[\widehat{\mathbf{p}}, H]\psi(\mathbf{r}, t)dx \qquad (3.10)$$

Proceeding in the same way as in the previous case, it is easy to see that

$$\langle[\hat{\mathbf{p}}, H]\rangle = i\hbar\frac{d}{dt}\int \psi^*(\mathbf{r}, t)\hat{\mathbf{p}}\psi(\mathbf{r}, t)$$

$$= i\hbar\frac{d}{dt}\langle\hat{\mathbf{p}}\rangle$$

Therefore from Eq. (3.9), we get

$$\frac{d}{dt}\langle\hat{\mathbf{p}}\rangle = -\langle\nabla V(\mathbf{r})\rangle$$

Q3.4 Using the result

$$[H, x] = -i\hbar\frac{\hat{p}}{m}, \tag{3.11}$$

for a particle of mass m moving in x-direction in a potential $V(x)$, show that the average value of its momentum in a stationary state with discrete energy is zero.

Solution: Suppose u_{E_n} is a stationary state of H with the discrete energy eigenvalue E_n so that

$$Hu_{E_n}(x) = E_n u_{E_n}(x)$$

Using Eq. (3.11), the average value of $\hat{p}$ is given by

$$\langle\hat{p}\rangle = \frac{im}{\hbar}\int u_{E_n}^*(x)[H, x]u_{E_n}(x)dx$$

$$= \frac{im}{\hbar}\left\{\int u_{E_n}^*(x)Hxu_{E_n}(x)dx\right.$$

$$\left. - \int u_{E_n}^*(x)xHu_{E_n}(x)dx\right\}$$

Since H is hermitian, we can write above equation as

$$\langle\hat{p}\rangle = \frac{im}{\hbar}\left\{\int(Hu_E(x))^*xu_E(x)dx\right.$$

$$\left. - \int u_E^*(x)x(Hu_E(x))dx\right\}$$

Then using the eigenvalue equation given above (E_n is a real number):

$$\langle \hat{p} \rangle = \frac{im}{\hbar} \left(E_n \int (u_{E_n}(x))^* x u_{E_n}(x) dx \right.$$

$$\left. - E_n \int u_{E_n}^*(x) x (u_{E_n}(x)) dx \right)$$

$$= 0$$

Q3.5 A particle is in a state

$$\psi(x) = \frac{1}{\sqrt{a}} \sin \left(\frac{5\pi x}{a} \right), \quad |x| \leq a$$

$$= 0 \quad \text{elsewhere,}$$

show that the probability for the particle to be found with momentum p is given by

$$\frac{100\pi a}{2\hbar} \frac{\sin^2(pa/\hbar)}{(p^2 a^2/\hbar^2 - 25\pi^2)}.$$

Solution: The required probability is $|C(p)|^2$, where

$$C(p) = \frac{1}{\sqrt{2\pi\hbar}} \int_{-a}^{a} e^{-i\frac{px}{\hbar}} \frac{1}{\sqrt{a}} \sin \left(\frac{5\pi x}{a} \right) dx$$

Changing $x \rightarrow -x$ in the integral,

$$C(p) = -\frac{1}{\sqrt{2\pi\hbar}} \int_{-a}^{a} e^{i\frac{px}{\hbar}} \frac{1}{\sqrt{a}} \sin \left(\frac{5\pi x}{a} \right) dx$$

Thus we can write

$$C(p) = -\frac{1}{\sqrt{2\pi\hbar}} \int_{-a}^{a} \left(\frac{e^{i\frac{px}{\hbar}} - e^{-i\frac{px}{\hbar}}}{2} \right) \frac{1}{\sqrt{a}} \sin \left(\frac{5\pi x}{a} \right) dx$$

$$= \frac{-i}{\sqrt{2\pi\hbar}} \int_{-a}^{a} \sin \left(\frac{px}{\hbar} \right) \frac{1}{\sqrt{a}} \sin \left(\frac{5\pi x}{a} \right) dx$$

Using the identity,

$$2 \sin \alpha \sin \beta = \cos (\alpha - \beta) - \cos (\alpha + \beta) \tag{3.12}$$

$$C(p) = \frac{-i}{\sqrt{2\pi\hbar a}} \frac{1}{2} \int_{-a}^{a} \left[\cos \left(\frac{px}{\hbar} - \frac{5\pi x}{a} \right) \right.$$
$$\left. - \cos \left(\frac{px}{\hbar} + \frac{5\pi x}{a} \right) \right] dx$$

$$= \frac{-i}{\sqrt{2\pi\hbar a}} \left[\frac{\sin \left(\frac{pa}{\hbar} - 5\pi \right)}{\left(\frac{p}{\hbar} - \frac{5\pi}{a} \right)} - \frac{\sin \left(\frac{pa}{\hbar} + 5\pi \right)}{\left(\frac{p}{\hbar} + \frac{5\pi}{a} \right)} \right]$$

Using

$$\sin(\alpha \pm \beta) = \sin \alpha \cos \beta \pm \cos \alpha \sin \beta \tag{3.13}$$

and the fact that $\sin 5\pi = 0$, $\cos 5\pi = -1$ we get

$$C(p) = \frac{i}{\sqrt{2\pi\hbar a}} \left[\frac{\sin \left(\frac{pa}{\hbar} \right)}{\left(\frac{p}{\hbar} - \frac{5\pi}{a} \right)} - \frac{\sin \left(\frac{pa}{\hbar} \right)}{\left(\frac{p}{\hbar} + \frac{5\pi}{a} \right)} \right]$$

$$= \frac{10\pi i}{a\sqrt{2\pi a\hbar}} \frac{a^2}{\frac{p^2 a^2}{\hbar^2} - 25\pi^2} \sin \left(\frac{pa}{\hbar} \right)$$

Thus the required probability is

$$|C(p)|^2 = \frac{100\pi a}{2\hbar} \frac{\sin^2 \left(\frac{pa}{\hbar} \right)}{\left(\frac{p^2 a^2}{\hbar^2} - 25\pi^2 \right)^2}$$

Q3.6 A particle of mass m is confined by an infinite square well potential $V(x) = 0, |x| \leq a; V(x) = \infty, |x| > a$. If the particle is in the state

$$\psi(x) = x, \quad |x| \leq a,$$
$$\psi(x) = 0, \quad |x| > a,$$

find the probability that a measurement of energy will give the result

$$E_n = \frac{\hbar^2}{2m} \left(\frac{\pi^2}{4a^2} \right) n^2.$$

Solution: Eigenfunctions corresponding to the eigenvalue E_n is

$$u_n(x) = \left(\frac{1}{a}\right)^{1/2} \cos\left(\frac{n\pi x}{2a}\right), \quad n \text{ odd}$$

Probability that a measurement of energy will give the result E_n is

$$P_{E_n} = \left|\int u_n^*(x)\psi(x)dx\right|^2$$

Now

$$\int u_n^*(x)\psi(x) = \left(\frac{1}{a}\right)^{1/2} \int_{-a}^{a} \cos\left(\frac{n\pi x}{2a}\right)x\,dx$$

Integration by parts gives

$$\int u_n^*(x)\psi(x) = \left(\frac{1}{a}\right)^{1/2} \left[x\left(\frac{2a}{n\pi}\sin\frac{n\pi x}{2a}\right)\Big|_{-a}^{+a}\right.$$

$$\left. -\int_{-a}^{a}\frac{2a}{n\pi}\sin\frac{n\pi x}{2a}dx\right]$$

$$= \left(\frac{1}{a}\right)^{1/2} \left[\frac{4a^2}{n\pi}(-1)^n + \left(\frac{2a}{n\pi}\right)^2 \cos\frac{n\pi x}{2a}\Big|_{-a}^{+a}\right]$$

$$= \frac{4a^{3/2}}{n\pi} \tag{3.14}$$

where we have used $\sin\frac{n\pi}{2} = (-1)^{n+1}$, and $\cos\frac{n\pi}{2} = 0$. Thus the required probability is

$$P_{E_n} = \left|\int u_n^*(x)\psi(x)dx\right|^2 = \frac{16a^3}{n^2}\pi^2$$

Note: We have taken $u_n(x)$ corresponding to even parity solution; the same result follows if we take odd parity solution:

$$u_n = \left(\frac{1}{a}\right)^{1/2} \sin\frac{n\pi x}{2a}, \quad n \text{ even}$$

Q3.7 For a free particle, find a wave function which is a simultaneous eigenfunction of both momentum and energy. This is not true for a particle moving in a potential since then $|\hat{p}, H| \neq 0$ [cf. Problem 3.3].

Solution: For a free particle, potential $V(\mathbf{r}) = 0$, and

$$H = \frac{\hat{\mathbf{p}}^2}{2m}$$

Let $u_{\mathbf{p}}(\mathbf{r})$ be eigenfunction of $\hat{\mathbf{p}}$ with eigenvalue $\mathbf{p}$

$$\hat{\mathbf{p}} u_{\mathbf{p}}(\mathbf{r}) = \mathbf{p} u_{\mathbf{p}}(\mathbf{r}) \tag{3.15}$$

Thus

$$H u_{\mathbf{p}}(\mathbf{r}) = \frac{\hat{\mathbf{p}}^2}{2m} u_{\mathbf{p}}(\mathbf{r}) = \frac{\hat{\mathbf{p}} \cdot \hat{\mathbf{p}}}{2m} u_{\mathbf{p}}(\mathbf{r})$$

$$= \frac{\mathbf{p}^2}{2m} u_{\mathbf{p}}(\mathbf{r}) = E u_{\mathbf{p}}(\mathbf{r})$$

Q3.8 (a) If a particle is in a state

$$\psi(x) = \left(\frac{2}{\pi \delta^2}\right)^{\frac{1}{4}} e^{-\frac{x^2}{\delta^2}}$$

find the probability of finding it in the momentum eigenstate

$$u_p(x) = \sqrt{\frac{1}{2\pi \hbar}} e^{-\frac{i}{\hbar} p x}.$$

(b) If it is in a state

$$u_p(x) = \sqrt{\frac{2}{\pi}} \sin nx, \quad 0 \leq x \leq \pi$$

show that the probability of finding it with momentum p is given by

$$|\phi(p)|^2 = \frac{1}{\pi^2 \hbar} \frac{\sin^2 \left(\left(\frac{p}{\hbar} - n\right)\frac{\pi}{2}\right)}{\left(\frac{p^2}{\hbar^2} - n^2\right)^2}.$$

Solution: (a) Required probability is $|\phi(p)^2$, where

$$\phi(p) = \frac{1}{\sqrt{2\pi\hbar}} \int_{-\infty}^{+\infty} e^{i\frac{px}{\hbar}} \psi(x) dx$$

$$= \frac{1}{\sqrt{2\pi\hbar}} \int_{-\infty}^{+\infty} e^{i\frac{px}{\hbar}} \left(\frac{2}{\pi\delta^2}\right)^{\frac{1}{4}} e^{-\frac{x^2}{\delta^2}} dx$$

$$= \frac{1}{\sqrt{2\pi\hbar}} \left(\frac{2}{\pi\delta^2}\right)^{\frac{1}{4}} e^{-\frac{\delta^2 p^2}{4\hbar^2}} \int_{-\infty}^{+\infty} e^{-\frac{1}{\delta^2}X^2} dX,$$

$$= \left(\frac{\delta^2}{2\pi\hbar^2}\right)^{1/4} e^{-\frac{\delta^2 p^2}{4\hbar^2}}$$

where $X = x - i\frac{\delta^2}{2\hbar}$.
Hence

$$|\phi(p)|^2 = \left(\frac{\delta^2}{2\hbar^2\pi}\right)^{1/2} e^{-\frac{\delta^2 p^2}{2\hbar^2}}$$

(b) Here

$$\phi(p) = \frac{1}{\sqrt{2\pi\hbar}} \int_0^\pi e^{i\frac{px}{\hbar}} \sqrt{\frac{2}{\pi}} \sin nx \, dx$$

$$= \frac{1}{\sqrt{\pi^2\hbar}} \frac{1}{2i} \int_0^\pi e^{i\frac{px}{\hbar}} (e^{inx} - e^{-inx}) dx$$

or

$$= \frac{1}{\sqrt{\pi^2\hbar}} \frac{1}{2i} \left[\frac{e^{i(n+p/\hbar)x}}{i(n + p/\hbar)} - (n \to -n)\right]_0^\pi$$

$$= -\frac{1}{\sqrt{\pi^2\hbar}} \left[e^{i\left(\frac{p}{\hbar}-n\right)\pi} \left(\frac{1}{n + p/\hbar} + \frac{1}{n - p/\hbar}\right)\right.$$

$$\left. - \left(\frac{1}{n + p/\hbar} + \frac{1}{n - p/\hbar}\right)\right]$$

$$= -\frac{n}{\sqrt{\pi^2\hbar}} \frac{1}{\frac{p^2}{\hbar^2} - n^2} \left[1 - e^{i\left(\frac{p}{\hbar}-n\right)\pi}\right] \tag{3.16}$$

where we have used

$$e^{-in\pi} = (-1)^n = e^{in\pi}$$

This gives

$$|\phi(p)|^2 = \frac{1}{\pi^2 \hbar} \frac{n^2}{\left(\frac{p^2}{\hbar^2} - n^2\right)^2} \left[1 + 1 - 2\mathrm{Re}\left(e^{i\left(\frac{p}{\hbar} - n\right)\pi}\right)\right]$$

$$= \frac{1}{\pi^2 \hbar} \frac{n^2}{\left(\frac{p^2}{\hbar^2} - n^2\right)^2} \left[2 - 2\cos\left(\frac{p}{\hbar} - n\right)\pi\right]$$

$$= \frac{1}{\pi^2 \hbar} \frac{4n^2}{\left(\frac{p^2}{\hbar^2} - n^2\right)^2} \sin^2\left(\left(\frac{p}{\hbar} - n\right)\frac{\pi}{2}\right)$$

Chapter 4

Solution of Problems
in Quantum Mechanics

Q4.1 A particle of mass m, moves in a potential

$$V(x) = \infty \quad x \leq 0$$
$$V(x) = 0 \quad 0 < x < a$$
$$V(x) = V \quad x \geq a$$

For $E < V$, find the condition for allowed energy values.

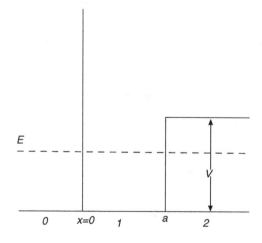

Solution: Define

$$k^2 = \frac{2mE}{\hbar} \quad \text{for regions 0 and 1}$$

$$K^2 = \frac{2m(V - E)}{\hbar} \quad \text{for region 2}$$

The Schrödinger equations for regions 1 and 2 can be written as

$$\left(\frac{d^2}{dx^2} + k^2\right)\psi_1(x) = 0$$

$$\left(\frac{d^2}{dx^2} - K^2\right)\psi_2(x) = 0$$

The following are solutions:

$$\psi_1(x) = Ae^{ikx} + Be^{-ikx}$$

$$\psi_2(x) = Ce^{-Kx}$$

Because the particle cannot go in the region 0 as V is ∞ at $x = 0$, $\psi_1(0) = 0$, which implies

$$A + B = 0; \quad A = -B \equiv \frac{F}{2i} \qquad (4.1)$$

Thus

$$\psi_1(x) = F \sin kx \qquad (4.2)$$

Now the boundary conditions at $x = a$ are:

$$\psi_1(a) = \psi_2(a)$$
$$\psi_1'(a) = \psi_2'(a) \qquad (4.3)$$

From Eqs. (4.2) and (4.3), we get

$$F \sin ka = Ce^{-Ka}$$
$$Fk \cos ka = -CKe^{-Ka}$$

Hence, we get

$$\tan ka = -\frac{k}{K}$$

the required condition for allowed energy values.

Q4.2 A particle of mass m approaches a potential barrier

$$V(x) = 0 \quad x < 0$$
$$= V \quad 0 \le x \le a$$
$$= 0 \quad x > a$$

from $x = -\infty$. For $E < V$, determine reflection and transmission coefficients for $Ka \ll 1$, where

$$K^2 = \frac{2m}{\hbar^2}(V - E).$$

Solution: Define

$$k^2 = \frac{2mE}{\hbar^2}$$

The Schrödinger equations for the three regions are

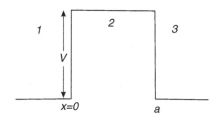

$$\left(\frac{d^2}{dx^2} + k^2\right)\psi_{1,3} = 0,$$

$$\left(\frac{d^2}{dx^2} - K^2\right)\psi_2 = 0$$

The following are solutions

$$\psi_1(x) = \underbrace{A\,e^{ikx}}_{\text{(incident)}} + \underbrace{B\,e^{-ikx}}_{\text{(reflected)}}$$

$$\psi_2(x) = C e^{Kx} + D e^{-Kx}$$

$$\psi_3(x) = \underbrace{F\,e^{ik(x-a)}}_{\text{(transmitted)}}$$

Boundary conditions at $x = 0$ and $x = a$, respectively give

$$A + B = C + D$$

$$ik(A - B) = K(C - D) \qquad (4.4)$$

$$Ce^{Ka} + De^{-Ka} = F$$

$$K(Ce^{Ka} - De^{-Ka}) = ikF \qquad (4.5)$$

Now for $Ka \ll 1$

$$e^{\pm Ka} \approx 1 \pm Ka$$

so that from Eq. (4.5)

$$(C + D) + Ka(C - D) = F$$

$$(C - D) + Ka(C + D) = i\frac{k}{K}F, \qquad (4.6)$$

which on using Eq. (4.4) give

$$(A + B) + ika(A - B) = F \qquad (4.7)$$

$$i\frac{k}{K}(A - B) + Ka(A + B) = i\frac{k}{K}F \qquad (4.8)$$

Solving these equations for B and F in terms of A, we obtain

$$B = -\frac{iAa(k^2 + K^2)}{2k - ia(k^2 - K^2)}$$

$$F = -\frac{2kA(1 - K^2a^2)}{2k - ia(k^2 - K^2)}$$

$$\approx -A\frac{2k}{2k - ia(k^2 - K^2)} \qquad (4.9)$$

Hence the Reflectivity:

$$R = \frac{|B|^2}{|A|^2} = \frac{a^2(k^2 + K^2)^2}{4k^2 + a^2(k^2 - K^2)^2}$$

and the Transmittivity:

$$T = \frac{|F|^2}{|A|^2} = \frac{4k^2}{4k^2 + a^2(k^2 - K^2)^2}$$

Q4.3 Consider a particle of mass m in a potential

$$
\begin{aligned}
V &= 0 & x < 0 \\
&= -V & 0 \leq x \leq a \\
&= 0 & x > a
\end{aligned}
$$

If $V \to \infty$ and $a \to 0$, such that $Va = \mu$, find the energy for the bound state $(E < 0)$.

Solution: Energy eigenvalue equation is

$$\left[-\frac{\hbar^2}{2m} \frac{d^2}{dx^2} + V(x) \right] \psi = E\psi$$

For bound states $|E| < V_0$, $E < 0$, $\epsilon = -E$. Define

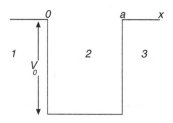

$$K^2 = -\frac{2mE}{\hbar^2} = \frac{2m\epsilon}{\hbar^2}$$

$$k^2 = \frac{2m}{\hbar^2}(E - (-V_0)) = \frac{2m}{\hbar^2}(V_0 - \epsilon)$$

The Schrödinger equations in the three regions are

$$\left(\frac{d^2}{dx^2} - K^2 \right) \psi_{1,3} = 0,$$

$$\left(\frac{d^2}{dx^2} + k^2 \right) \psi_2 = 0$$

where

$$\psi_1 = Ae^{Kx} = Ae^{-K|x|} \quad e^{-Kx} \text{ not possible for } x < 0$$
$$\psi_3 = De^{-Kx} \quad\quad\quad\quad e^{Kx} \text{ not possible for } x > 0$$
$$\psi_2 = Be^{ikx} + Ce^{-ikx}$$

Boundary conditions are

$$\psi_1(0) = \psi_2(0), \quad \psi_2(a) = \psi_3(a)$$
$$\psi_1'(0) = \psi_2'(0) \quad \psi_2'(a) = \psi_3'(a)$$

By applying these boundary conditions we get

$$A = B + C \quad\quad\quad Be^{ika} + Ce^{-ika} = De^{-Ka}$$
$$KA = ik(B - C) \quad ik(Be^{ika} - Ce^{-ika}) = -KDe^{-Ka}$$

Out of four constants, one arbitrary constant is to be fixed by normalization conditions. We have 3 constants but 4 relations to fix them. From the first two of the above conditions

$$B = \frac{A}{2}\left(1 + \frac{K}{ik}\right), \quad C = \frac{A}{2}\left(1 - \frac{K}{ik}\right)$$

Then substituting in the other 2 conditions, we get the following relation on energy eigenvalues:

$$\frac{\sin ka - \frac{K}{k}\cos ka}{\cos ka + \frac{K}{k}\sin ka} = \frac{K}{k}$$

or

$$(k\sin ka - K\cos ka) = K\cos ka + \frac{K^2}{k}\sin ka \quad (4.10)$$

Now $V_0 \to \infty$, $a \to 0$, such that $V_0 a = \mu$

$$k^2 a \to \frac{2m\mu}{\hbar^2}, \quad \text{as } k \to \infty, a \to 0$$

$$ka = a^{1/2}\sqrt{\frac{2m\mu}{\hbar^2}} \quad \to 0$$

$$\sin ka \approx ka, \quad \cos ka \to 1$$

From Eq. (4.10), keeping first order term in $ka^{1/2}$ or k^2a

$$k(ka) - K = K + \frac{K^2}{k}(ka)$$

$$= K + K(Ka) = K$$

This gives

$$k^2a = 2K$$

i.e.

$$\frac{2m\mu}{\hbar^2} = 2\left(\frac{2m\epsilon}{\hbar^2}\right)^{1/2}$$

$$\epsilon = \frac{m}{2\hbar^2}\mu^2$$

$$E = -\frac{m}{2\hbar^2}\mu^2 \tag{4.11}$$

giving the energy of the bound state.

Q4.4 Consider a rectangular potential barrier 4 volts high and 10^{-9} meters wide. Calculate a rough value for the probability for an electron of kinetic energy 3 volts to penetrate the barrier.

Solution: Penetration probability is given by [c.f. Eq. (4.2) of the text]

$$T(E_0) = 16\frac{E_0}{V}\left(1 - \frac{E_0}{V}\right)e^{\left[-\sqrt{\frac{8m}{\hbar^2}(V-E_0)}a\right]}$$

$$a = 10^{-9}m$$

$$V = 4\text{eV}$$

$$E_0 = 3\text{eV}$$

By using these numerical values we can find

$$\sqrt{\frac{8m}{\hbar^2}(V - E_0)} \approx 10^{10}m^{-1}$$

and

$$\sqrt{\frac{8m}{\hbar^2}(V - E_0)}a \approx 10^{10} \times 10^{-9} = 10$$

so

$$T(E_0) \approx 3(e^{-10}) \approx 10^{-4}$$

Q4.5 A particle of mass m, moves in a potential (with energy $E < V$)

$$V(x) = \infty \quad x \leq 0$$
$$V(x) = 0 \quad 0 < x < a$$
$$V(x) = V \quad a \leq x \leq b$$
$$V(x) = 0 \quad x > b$$

and the energy has a value as obtained in Problem 4.1, find the relative intensity at $x = b$ and $x = a$. For the region $x > b$, show that there is equal intensity in the beams travelling to left and right.

Solution: Define

$$k^2 = \frac{2mE}{\hbar^2}$$

$$K^2 = \frac{2m(V-E)}{\hbar^2} \quad E < V$$

Now

$$\psi_1(x) = 0 \quad x \leq 0.$$

Schrödinger equations in various regions are

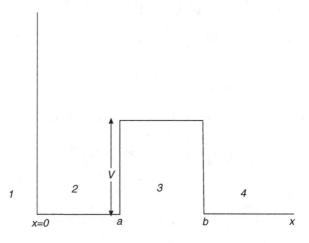

$$\left(\frac{d^2}{dx^2} + k^2\right)\psi_{2,4} = 0$$

$$\left(\frac{d^2}{dx^2} - K^2\right)\psi_3 = 0$$

and have solutions

$$\psi_2(x) = A'e^{ikx} + B'e^{-ikx} \qquad 0 < x < a$$
$$\psi_3(x) = Ce^{K(x-a)} + De^{-K(x-a)} \qquad a < x < b$$
$$\psi_4(x) = Fe^{ik(x-b)} + Ge^{-ik(x-b)} \qquad x > b$$

Now

$$\psi_1(0) = \psi_2(0)$$

$$A' + B' = 0 \Rightarrow B' = -A' \equiv \frac{A}{2i}$$

therefore,

$$\psi_2(x) = A\sin kx \quad 0 < x < a$$

Boundary conditions at $x = a$ and $x = b$ give, respectively

$$\left.\begin{array}{l} A\sin ka = C + D \\ Ak\cos ka = K(C - D) \end{array}\right] \Rightarrow \begin{array}{l} C = \dfrac{A}{2}\left[\sin ka + \dfrac{k}{K}\cos ka\right] \\[2mm] D = \dfrac{A}{2}\left[\sin ka - \dfrac{k}{K}\cos ka\right] \end{array}$$

$$(4.12)$$

$$\left.\begin{array}{l} Ce^{K(b-a)} + De^{-K(b-a)} = F + G \\ K[ce^{k(b-a)} - De^{-K(b-a)}] = ik[F - G] \end{array}\right] \Rightarrow$$

$$F = \frac{A}{2}\left[\left(1 + \frac{K}{ik}\right)Ce^{K(b-a)} + \left(1 - \frac{K}{ik}\right)De^{-K(b-a)}\right]$$

$$G = \frac{A}{2}\left[\left(1 - \frac{K}{ik}\right)Ce^{K(b-a)} + \left(1 + \frac{K}{ik}De^{-K(b-a)}\right)\right]$$

$$(4.13)$$

so

$$T(E_0) \approx 3(e^{-10}) \approx 10^{-4}$$

Q4.5 A particle of mass m, moves in a potential (with energy $E < V$)

$$V(x) = \infty \quad x \leq 0$$
$$V(x) = 0 \quad 0 < x < a$$
$$V(x) = V \quad a \leq x \leq b$$
$$V(x) = 0 \quad x > b$$

and the energy has a value as obtained in Problem 4.1, find the relative intensity at $x = b$ and $x = a$. For the region $x > b$, show that there is equal intensity in the beams travelling to left and right.

Solution: Define

$$k^2 = \frac{2mE}{\hbar^2}$$

$$K^2 = \frac{2m(V - E)}{\hbar^2} \quad E < V$$

Now

$$\psi_1(x) = 0 \quad x \leq 0.$$

Schrödinger equations in various regions are

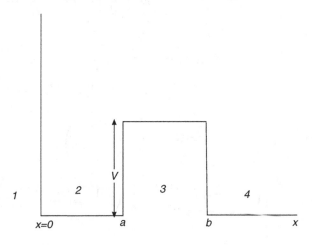

Fig. 4.7. Show that (i) for $K \neq k$, $T = 1$ and $R = 0$ occur for energies E given by

$$E = -V + \frac{n^2\pi^2\hbar^2}{8ma^2}$$

(ii) T has a minimum value for energies

$$E = -V + \frac{(2n+1)^2\pi^2\hbar^2}{32ma^2}$$

(iii) as $E \to \infty$, $K \to k$, $T \to 1$.

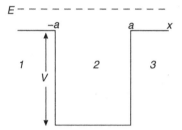

Solution: Define

$$k^2 = \frac{2mE}{\hbar^2}, \qquad K^2 = \frac{2m(E+V)}{\hbar^2}$$

The Schrödinger equations for regions 1, 3 and 2 can be written as

$$\left(\frac{d^2}{dx^2} + k^2\right)\psi_{1,3}(x) = 0$$

$$\left(\frac{d^2}{dx^2} + K^2\right)\psi_2(x) = 0$$

which have the solutions

$$\psi_1(x) = \underbrace{A\,e^{ikx}}_{\text{(incident)}} + \underbrace{B\,e^{-ikx}}_{\text{(reflected)}}$$

$$\psi_2(x) = C e^{iKx} + D e^{-iKx}$$

$$\psi_3(x) = \underbrace{F\,e^{ikx}}_{\text{(transmitted)}}$$

The boundary condition at $x = -a$ implies

$$\psi_1(-a) = \psi_2(-a) \Rightarrow Ae^{-ika} + Be^{ika} = Ce^{iKa} + De^{iKa}$$

$$\psi_1'(-a) = \psi_2'(-a) \Rightarrow ik(Ae^{-ika} - Be^{ika})$$
$$= iK(Ce^{iKa} - De^{iKa})$$

By addition and subtraction of these equations we can find A and B in terms of C and D

$$A = \frac{e^{ika}}{2}\left(1 + \frac{k}{K}\right)Ce^{-iKa} + \left(1 - \frac{k}{K}\right)De^{iKa} \quad (4.16)$$

$$B = \frac{e^{-ika}}{2}\left(1 - \frac{k}{K}\right)Ce^{-iKa} + \left(1 + \frac{k}{K}\right)De^{iKa}$$

Similarly, the boundary condition at $x = a$ gives

$$\psi_2(a) = \psi_3(a) \Rightarrow Ce^{iKa} + De^{-iKa} = Fe^{ika}$$

$$\psi_2'(a) = \psi_3'(a) \Rightarrow iK(Ce^{iKa} - De^{-iKa}) = ikFe^{ika}$$

which give

$$C = \frac{1}{2}e^{-iKa}F\left(1 + \frac{k}{K}\right)e^{ika}$$

$$D = \frac{1}{2}e^{iKa}F\left(1 - \frac{k}{K}\right)e^{ika}$$

Putting these values of C and D in Eq. (4.16), we get

$$A = \frac{F}{4}e^{2ika}\left[4\cos 2Ka - 2i\left(\frac{K}{k} + \frac{k}{K}\right)\sin 2Ka\right]$$

or

$$F = \frac{2Ae^{-2ika}}{[2\cos 2Ka - i(\frac{K}{k} + \frac{k}{K})\sin 2Ka]}$$

Similarly,

$$B = \frac{F}{4}\left[2i\left(\frac{K}{k} - \frac{k}{K}\right)\sin 2Ka \right]$$

$$= \frac{2i}{4}\frac{2Ae^{-2ika}\left(\frac{K}{k} - \frac{k}{K}\right)\sin 2Ka}{2\cos 2Ka - i\left(\frac{K}{k} + \frac{k}{K}\right)\sin 2Ka}$$

Transmittivity:

$$T = \frac{|F|^2}{|A|^2} = \frac{4}{4\cos^2 2Ka + (\frac{K}{k} + \frac{k}{K})^2 \sin^2 2Ka} \tag{4.17}$$

Reflectivity:

$$R = \frac{|B|^2}{|A|^2} = \frac{(\frac{K}{k} - \frac{k}{K})^2 \sin^2 2Ka}{4\cos^2 2Ka + (\frac{K}{k} + \frac{k}{K})^2 \sin^2 2Ka}$$

and

$$T + R = \frac{4 + (\frac{K}{k} - \frac{k}{K})^2 \sin^2 2Ka}{4\cos^2 2Ka + (\frac{K}{k} + \frac{k}{K})^2 \sin^2 2Ka}$$

$$= \frac{4 + (\frac{K}{k} + \frac{k}{K})^2 \sin^2 2Ka - 4\sin^2 2Ka}{4\cos^2 2Ka + (\frac{K}{k} + \frac{k}{K})^2 \sin^2 2Ka}$$

$$= 1$$

(i) From Eq. (4.17) transmittivity can be written in the following form

$$T = \frac{1}{1 + \frac{1}{4}\frac{(K^2 - k^2)^2}{k^2 K^2}\sin^2 2Ka}$$

Putting the values of k and K into above equation, we get

$$T = \frac{1}{\left[1 + \frac{V^2 \sin^2 2Ka}{4E(E+V)}\right]}$$

$$= \frac{(4\frac{E}{V})(1 + \frac{E}{V})}{4\frac{E}{V}(1 + \frac{E}{V}) + \sin^2 2Ka}$$

This implies that $T = 1$ and $R = 0$ occur when

$$\sin 2Ka = 0 \Rightarrow 2Ka = n\pi$$

$$4K^2a^2 = n^2\pi^2$$

or

$$2m(E + V) = \frac{n^2\pi^2\hbar^2}{4a^2} \qquad (4.18)$$

i.e. when

$$E = -V + \frac{n^2\pi^2\hbar^2}{8ma^2}$$

(ii) Similarly, T has a minimum value when

$$\sin 2Ka = 1 \Rightarrow 2Ka = (2n + 1)\frac{\pi}{2}$$

$$4K^2a^2 = (2n + 1)^2\frac{\pi^2}{4}$$

$$\frac{2m(E + V)}{\hbar^2} = (2n + 1)^2\frac{\pi^2}{4}$$

Hence

$$E = -V + \frac{\hbar^2(2n + 1)^2\pi^2}{32ma^2}$$

(iii) as $E \to \infty$, $k \to K$, it means that transmittivity becomes

$$T \to \frac{4}{4\cos^2 2Ka + (2)^2\sin^2 2Ka}$$

$$= 1$$

Chapter 5

Simple Harmonic Oscillator

The following formulas are relevant for the solution of problems 5.1 to 5.6.

From Eqs. (5.13), (5.14), (5.17), (5.39) and (5.40) of the text:

$$x = \sqrt{\frac{\hbar}{2m\omega}}(a + a^\dagger)$$

$$\hat{p} = -i\sqrt{\frac{m\hbar\omega}{2}}(a - a^\dagger)$$

$$H = \frac{\hbar\omega}{2}(aa^\dagger + a^\dagger a)$$

$$(u_n|u_m) = \delta_{mn}$$

Q5.1 Show that

$$\langle x \rangle_n = \int u_n^*(x)xu_n(x)dx = (u_n|xu_n) = 0.$$

$$\langle p \rangle_n = \int u_n^*(x)\left(-i\hbar\frac{\partial}{\partial x}\right)u_n(x)dx = (u_n|\hat{p}u_n) = 0 \ .$$

Solution: Using above equations

$$\langle x \rangle = (u_n|xu_n)$$

$$= \sqrt{\frac{\hbar}{2m\omega}}[\sqrt{n}(u_n|u_{n-1}) + \sqrt{n+1}(u_n|u_{n+1})]$$

$$= 0$$

$$\langle \hat{p} \rangle = (u_n | \hat{p} u_n)$$

$$= -i\sqrt{\frac{m\hbar\omega}{2}} \left[\sqrt{n}(u_n | u_{n-1}) - \sqrt{n+1}(u_n | u_{n+1}) \right]$$

$$= 0$$

Q5.2 Show that

$$(u_n | a^2 u_n) = 0$$

$$(u_n | a^{\dagger^2} u_n) = 0.$$

Solution:

$$a^2 u_n = a(\sqrt{n} u_{n-1})$$

$$= \sqrt{n(n-1)} u_{n-2}$$

similarly

$$a^{\dagger 2} u_n = \sqrt{(n+1)(n+2)} u_{n+2} \qquad (5.1)$$

Hence

$$(u_n | a^2 u_n) = 0 = (u_n | a^{\dagger 2} u_n) \qquad (5.2)$$

Q5.3 Show that for a simple harmonic oscillator

$$\langle V \rangle_n = \frac{1}{2} \left(n + \frac{1}{2} \right) \hbar\omega$$

$$\langle T \rangle_n = \frac{1}{2} \left(n + \frac{1}{2} \right) \hbar\omega$$

$$\langle T \rangle_n = \langle V \rangle_n \quad : \quad Virial \quad Theorm.$$

Hint: First show that

$$(u_n | (a + a^\dagger)^2 u_n) = (2n + 1)$$

$$(u_n | (a - a^\dagger)^2 u_n) = (2n + 1).$$

Solution: Now

$$x^2 = \frac{\hbar}{2m\omega}(a + a^\dagger)^2$$

$$= \frac{\hbar}{2m\omega}(a^2 + aa^\dagger + a^\dagger a + a^{\dagger 2})$$

$$= \frac{\hbar}{2m\omega}\left(\frac{2}{\hbar\omega}H + a^2 + a^{\dagger 2}\right) \tag{5.3}$$

$$\hat{p}^2 = \left(\frac{m\hbar\omega}{2}\right)\left(\frac{2H}{\hbar\omega} - a^2 - a^{\dagger 2}\right) \tag{5.4}$$

therefore

$$\langle V \rangle_n = \frac{1}{2}m\omega^2 \langle x^2 \rangle_n$$

$$= \frac{1}{2}m\omega^2 \frac{\hbar}{2m\omega}\frac{2}{\hbar\omega}\langle H \rangle_n$$

$$= \frac{1}{2}\langle H \rangle_n = \frac{1}{2}\left(n + \frac{1}{2}\right)\hbar\omega$$

$$\langle T \rangle_n = \frac{1}{2m}\langle \hat{p}^2 \rangle_n = \frac{1}{2m}\frac{m\hbar\omega}{2}\frac{2}{\hbar\omega}\langle H \rangle_n$$

$$= \frac{1}{2}\left(n + \frac{1}{2}\right)\hbar\omega \tag{5.5}$$

Hence

$$\langle T \rangle_n = \langle V \rangle_n : \quad \text{Virial theorem}$$

Q5.4 Show that for a simple harmonic oscillator

$$\Delta x \Delta p = \hbar\left(n + \frac{1}{2}\right)$$

$$\Delta x = [\langle (x - \langle x \rangle)^2 \rangle]^{\frac{1}{2}}$$

$$\Delta p = [\langle (p - \langle p \rangle)^2 \rangle]^{\frac{1}{2}}$$

Solution:

$$\Delta x = \left[\langle (x - \langle x \rangle)^2 \rangle\right]^{1/2}$$

$$= \left[\langle x^2 \rangle - 2\langle x \rangle \langle x \rangle + \langle x \rangle^2\right]^{1/2}$$

$$= \left[\langle x^2 \rangle\right]^{1/2}$$

$$= \left[\frac{1}{m\omega^2}(n + 1/2)\hbar\omega\right]^{1/2}$$

Similarly

$$\Delta p = \left[\langle p^2 \rangle\right]^{1/2}$$

$$= \left[m(n + 1/2)\hbar\omega\right]^{1/2}$$

In deriving the above results, we have used Eqs. (5.3) and (5.4).

Hence we have

$$\Delta x \Delta p = (n + 1/2)\hbar$$

In particular for the ground state

$$\Delta x \Delta p = \frac{\hbar}{2}$$

Q5.5 Show that

$$p_{nk} = -i\frac{\sqrt{m\omega\hbar}}{2}\left[\sqrt{n+1}\delta_{k,n+1} - \sqrt{n}\delta_{k,n-1}\right]$$

$$\langle p \rangle = -\sqrt{2m\omega\hbar}\sum_n \sqrt{n}|C_n||C_{n-1}|\sin(\omega t + \delta_{n-1} - \delta_n)$$

Further show that

$$\langle x \rangle_t = \langle x \rangle_0 + \frac{\langle p_0 \rangle}{m\omega}\sin\omega t$$

$$\langle p \rangle_t = \langle p \rangle_0 \cos\omega t - m\omega\langle x \rangle_0 \sin\omega t.$$

in complete correspondence with classical equations.

Solution: Now

$$\hat{p} = -i\hbar\frac{\partial}{\partial x} = -i\sqrt{\frac{m\omega\hbar}{2}}(a - a^\dagger)$$

$$\therefore p_{nk} = -i\sqrt{\frac{m\omega\hbar}{2}}[\sqrt{k}(u_n|u_{k-1}) - \sqrt{k+1}(u_n|u_{k+1})]$$

$$= -i\sqrt{\frac{m\omega\hbar}{2}}[\sqrt{k}\delta_{n,k-1} - \sqrt{k+1}\delta_{n,k+1}]$$

$$= -i\sqrt{\frac{m\omega\hbar}{2}}[\sqrt{n+1}\delta_{k,n+1} - \sqrt{n}\delta_{k,n-1}]$$

$$\langle p \rangle = \int \psi^*(x,t)\left(-i\hbar\frac{\partial}{\partial x}\right)\psi(x,t)dx$$

Now [cf Eq. (5.68) of the text]

$$\psi(x,t) = \sum_n C_n u_n(x)e^{-iE_n t/\hbar}$$

therefore

$$\langle p \rangle = \sum_n\sum_k C_n^* C_k e^{i(E_n-E_k)t/\hbar}\int u_n^*\left(-i\hbar\frac{\partial}{\partial x}\right)u_k dx$$

$$= \sum_n\sum_k C_n^* C_k e^{i(E_n-E_k)t/\hbar}(u_n|\hat{p}u_k)$$

$$= -i\sqrt{\frac{m\omega\hbar}{2}}\sum_n\sum_k[\sqrt{n+1}\delta_{k,n+1} - \sqrt{n}\delta_{k,n-1}]$$

$$\times C_n^* C_k e^{i(E_n-E_k)t/\hbar}$$

Now

$$E_n - E_k = \hbar\omega(n + 1/2 - k - 1/2)$$
$$= \hbar\omega(n - k)$$

Thus

$$\langle p \rangle = -i\sqrt{\frac{m\omega\hbar}{2}}\left[\sum_n C_n^* C_{n+1} e^{-i\omega t}\sqrt{n+1}\right.$$

$$\left. -\sum_n C_n^* C_{n-1} e^{i\omega t}\sqrt{n}\right]$$

$$= i\sqrt{\frac{m\omega\hbar}{2}}\left[\sum_n^\infty \sqrt{n}(C_n^* C_{n-1} e^{i\omega t} - C_{n-1}^* C_n e^{-i\omega t})\right]$$

Now we can write

$$C_n = |C_n|e^{i\delta_n}, \quad C_{n-1} = |C_{n-1}|e^{i\delta_{n-1}}$$

Hence

$$\langle p \rangle_t = i\sqrt{\frac{m}{2}}\sum_n \sqrt{n\hbar\omega}|C_n||C_{n-1}|$$

$$\times\left[e^{i(\omega t - \delta_n + \delta_{n-1})} - e^{-i(\omega t - \delta_n + \delta_{n-1})}\right]$$

$$= -\sqrt{2m}\sum_n \sqrt{n\hbar\omega}|C_n||C_{n-1}|$$

$$\times \sin\left(\omega t + \delta_{n-1} - \delta_n\right)$$

whereas

$$\langle x \rangle_t = \sqrt{\frac{2}{m\omega^2}}\sum_n \sqrt{n\hbar\omega}|C_n||C_{n-1}|\cos\left(\omega t - \delta_{n-1} - \delta_n\right)$$

Thus

$$\langle x \rangle_0 = \sqrt{\frac{2}{m\omega^2}}\sum_n \sqrt{n\hbar\omega}|C_n||C_{n-1}|\cos\left(\delta_{n-1} - \delta_n\right)$$

$$\langle p \rangle_0 = -\sqrt{2m}\sum_n \sqrt{n\hbar\omega}|C_n||C_{n-1}|\sin\left(\delta_{n-1} - \delta_n\right)$$

Hence

$$\langle x \rangle_t = \sqrt{\frac{2}{m\omega^2}} \sum_n \sqrt{n\hbar\omega} |C_n||C_{n-1}|[\cos \omega t \cos (\delta_{n-1} - \delta_n)$$

$$- \sin \omega t \sin (\delta_{n-1} - \delta_n)]$$

$$= \langle x \rangle_0 \cos \omega t + \sqrt{\frac{2}{m\omega^2}} \frac{\langle p_0 \rangle}{\sqrt{2m}} \sin \omega t$$

$$= \langle x \rangle_0 \cos \omega t + \frac{\langle p \rangle_0}{m\omega} \sin \omega t$$

$$\langle p \rangle_t = -\sqrt{2m} \sum_n \sqrt{n\hbar\omega} |C_n||C_{n-1}|[\sin \omega t \cos (\delta_{n-1} - \delta_n)$$

$$+ \cos \omega t \sin (\delta_{n-1} - \delta_n)]$$

$$= -\sqrt{2m} \sqrt{\frac{m\omega^2}{2}} \langle x \rangle_0 \sin \omega t + \langle p \rangle_0 \cos \omega t$$

$$= \langle p \rangle_0 \cos \omega t - m\omega \langle x_0 \rangle \sin \omega t$$

Q5.6 Verify that

$$\Psi(x,t) = \sqrt{\frac{2}{3}} \left[\left(\frac{\beta}{\sqrt{\pi}} \right)^{\frac{1}{2}} e^{\frac{-i\omega t}{2}} + \left(\frac{2\beta^3}{\sqrt{\pi}} \right)^{\frac{1}{2}} x e^{\frac{-i3\omega t}{2}} \right] e^{\frac{-\beta^2 x^2}{2}}$$

is a solution of time dependent Schrödinger equation for a particle of mass m in a simple harmonic oscillator potential. Calculate expectation values of x and p and show that

$$\frac{d\bar{x}}{dt} = \frac{\bar{p}}{m}.$$

Hint, note that 1st and 2nd term in $\Psi(x,0)$ are normalized eigenfunctions of the time independent Schrödinger equation $H\Psi = E\Psi$ with eigenvalues $\frac{\hbar\omega}{2}$ and $\frac{3\hbar\omega}{2}$ respectively.

Solution: Now

$$Hu_0(x) = \frac{1}{2}\hbar\omega u_0(x)$$

$$Hu_1(x) = \frac{3}{2}\hbar\omega u_1(x)$$

where

$$u_0(x) = \left(\frac{\beta}{\sqrt{\pi}}\right)^{1/2} e^{-\beta^2 x^2/2}$$

$$u_1(x) = \left(\frac{2\beta^3}{\sqrt{\pi}}\right)^{1/2} x e^{-\beta^2 x/2}, \quad \beta = \sqrt{\frac{m\omega}{\hbar}}$$

are normalized eigenfunctions of H with eigenvalues $\frac{1}{2}\hbar\omega$ and $\frac{3}{2}\hbar\omega$, respectively.

Thus

$$\psi(x,t) = \sqrt{2/3}\left[u_0(x)e^{-i\omega t/2} + u_1(x)e^{-i3\omega t/2}\right]$$

Now

$$i\hbar\frac{\partial\psi(x,t)}{\partial t} = \sqrt{2/3}i\hbar\left[-i\omega/2u_0(x)e^{-i\omega t/2}\right.$$

$$\left. -i3\omega/2u_1(x)e^{-i3\omega t/2}\right]$$

$$= \sqrt{2/3}\left[e^{-i\omega t/2}\frac{\hbar}{2}u_0(x) + \frac{3\hbar\omega}{2}e^{-i3\omega t/2}u_1(x)\right]$$

and

$$H\psi(x,t) = \sqrt{2/3}\left[e^{-i\omega t/2}\frac{\hbar\omega}{2}u_0(x) + \frac{3\hbar\omega}{2}e^{-i3\omega t/2}u_1(x)\right]$$

showing that the Schrödinger equation

$$i\hbar\frac{\partial\psi(x,t)}{\partial t} = H\psi(x,t)$$

is satisfied.

Hence $\psi(x,t)$ is a solution of the time dependent Schrödinger equation for a particle of mass m in a simple harmonic potential. Now

$$\bar{x} = \langle x \rangle = \int \psi^*(x, t) x \psi(x, t) dx$$

$$= 2/3 \int \left[u_0^* x u_0 + u_1^* x u_1 + e^{-i\omega t} u_0^* x u_1 \right.$$

$$\left. + e^{i\omega t} u_1^* x u_0 \right] dx$$

$$= \frac{2}{3} \left[(u_0|xu_0) + (u_1|xu_1) + e^{-i\omega t} (u_0|xu_1) \right.$$

$$\left. + e^{i\omega t} (u_1|xu_0) \right] \tag{5.6}$$

Similarly

$$\bar{p} = (p) = \frac{2}{3} \left[(u_0|\hat{p}u_0) + (u_1|\hat{p}u_1) + e^{-i\omega t} (u_0|\hat{p}u_1) \right.$$

$$\left. + e^{i\omega t} (u_1|\hat{p}u_0) \right] \tag{5.7}$$

Now

$$(u_n|xu_m) = \sqrt{\frac{\hbar}{2m\omega}} \left[(u_n|au_m) + (u_n|a^\dagger u_m) \right]$$

$$(u_n|\hat{p}u_m) = -i\sqrt{\frac{m\hbar\omega}{2}} \left[(u_n|au_m) - (u_n|a^\dagger u_m) \right]$$

therefore

$$(u_n|xu_m) = \sqrt{\frac{\hbar}{2m\omega}} \left[\sqrt{m}(u_n|u_{m-1}) + \sqrt{m+1}(u_n|u_{m+1}) \right]$$

$$(u_n|\hat{p}u_m) = -i\sqrt{\frac{m\hbar\omega}{2}} \left[\sqrt{m}(u_n|u_{m-1}) \right.$$

$$\left. - \sqrt{m+1}(u_n|u_{m+1}) \right]$$

Hence from Eqs. (5.6) and (5.7), using the above equations, we have

$$\bar{x} = \frac{2}{3} \left[e^{-i\omega t} + e^{i\omega t} \right] \sqrt{\frac{\hbar}{2m\omega}} = \frac{2}{3} \sqrt{\frac{\hbar}{2m\omega}} 2 \cos \omega t$$

$$\bar{p} = \frac{2}{3}(-i)\sqrt{\frac{m\hbar\omega}{2}} \left[e^{-i\omega t} - e^{i\omega t} \right] = -\frac{2}{3} \sqrt{\frac{m\hbar\omega}{2}} 2 \sin \omega t$$

Thus

$$\frac{d\bar{x}}{dt} = \frac{2}{3}\sqrt{\frac{\hbar}{2m\omega}}(-2\omega \sin \omega t)$$

$$= -\frac{2}{3}\sqrt{\frac{\hbar\omega}{2m}}2 \sin \omega t$$

$$= \frac{\bar{p}}{m}$$

Q5.7 A particle of mass m moves in a potential

$$V(x) = \frac{1}{2}m\omega^2 x^2 \quad x > 0$$

$$= \infty \quad\quad x < 0$$

find the energy eigenvalues.

Solution:

$$V(x) = \frac{1}{2}m\omega^2 x^2 \quad x > 0$$

$$= \infty \quad\quad x < 0$$

Only solutions for which $u_n(0) = 0$ are allowed.
Now the Hermite polynomial has the property

$$H_n(0) = 0 \quad \text{for odd } n$$

Thus only odd n solutions are allowed.
Hence the energy levels are given by

$$E_n \left[(2n-1) + \frac{1}{2}\right]\hbar\omega = \frac{1}{2}(4n-1)\hbar\omega \quad n = 1, 2, 3, 4 \cdots$$

viz.

$$\frac{3}{2}\hbar\omega, \frac{7}{2}\hbar\omega, \frac{11}{2}\hbar\omega \cdots$$

Q5.8 Show that the Schrödinger eqution for a simple harmonic
oscillator in momentum space is given by:

$$\frac{\partial^2}{\partial p^2}\phi(p) + \frac{2}{m\hbar^2\omega^2}\left(E - \frac{p^2}{2m}\right)\phi(p) = 0.$$

Solution: The Schrödinger equation for a simple harmonic oscillator is

$$\left[-\frac{\hbar^2}{2m} \frac{\partial^2}{\partial x^2} + \frac{1}{2} m\omega^2 x^2 \right] \psi(x) = E\psi(x) \qquad (5.8)$$

The wave function $\psi(x)$ in position space and the wave function $\phi(p)$ in momentum space are related to each other by Fourier transform, viz.

$$\psi(x) = \frac{1}{\sqrt{2\pi\hbar}} \int e^{ipx/\hbar} \phi(p) dp$$

$$\phi(p) = \frac{1}{\sqrt{2\pi\hbar}} \int e^{-ipx/\hbar} \psi(x) dx \qquad (5.9)$$

Now

$$\frac{\partial^2 \psi}{\partial x^2} = \frac{1}{\sqrt{2\pi\hbar}} \int \left(-\frac{p^2}{\hbar^2} \right) e^{ipx/\hbar} \phi(p) dp$$

Thus

$$\int e^{-ip'x/\hbar} \frac{\partial^2 \psi}{\partial x^2} dx$$

$$= \frac{1}{\sqrt{2\pi\hbar}} \left(-\frac{1}{\hbar^2} \right) \int p^2 \phi(p) dp \int e^{i(p-p')x/\hbar} dx$$

$$= \frac{1}{\sqrt{2\pi\hbar}} \left(-\frac{1}{\hbar^2} \right) \int p^2 \phi(p) dp (2\pi\hbar) \delta(p - p')$$

$$= -\frac{\sqrt{2\pi\hbar}}{\hbar^2} p'^2 \phi(p') \qquad (5.10)$$

therefore

$$\frac{1}{\sqrt{2\pi\hbar}} \int e^{-ipx/\hbar} \frac{\partial^2 \psi}{\partial x^2} dx = -\frac{1}{\hbar^2} p^2 \phi(p) \qquad (5.11)$$

From Eq. (5.9):

$$\frac{\partial^2 \phi(p)}{\partial p^2} = \frac{1}{\sqrt{2\pi\hbar}} \left(-\frac{1}{\hbar^2} \right) \int x^2 \psi(x) dx \qquad (5.12)$$

Now from Eq. (5.8)

$$\frac{1}{\sqrt{2\pi\hbar}} \int e^{-ipx/\hbar} \left[-\frac{\hbar^2}{2m}\frac{\partial^2\psi}{\partial x^2} + \frac{1}{2m}\omega^2 x^2 \psi(x) \right] dx$$

$$= \frac{1}{\sqrt{2\pi\hbar}} E \int e^{-ipx/\hbar}\psi(x)dx$$

Hence on using Eqs. (5.9), (5.11) and (5.12), we get

$$-\frac{\hbar^2}{2m}\left(-\frac{p^2}{\hbar^2}\right)\phi(p) + \frac{1}{2}m\omega^2(-\hbar^2)\frac{\partial^2\phi(p)}{\partial p^2} = E\phi(p)$$

or

$$\frac{\partial^2\phi(p)}{\partial p^2} + \frac{2}{m\hbar^2\omega^2}\left(E - \frac{p^2}{2m}\right)\phi(p) = 0$$

Q5.9 Obtain the ground state and first excited state wave functions $\phi_0(p)$ and $\phi_1(p)$ in momentum space

Answer

$$\phi_0(p) = \left(\frac{1}{\pi m\hbar\omega}\right)^{\frac{1}{4}} e^{-\frac{p^2}{2m\hbar\omega}}$$

$$\phi_1(p) = \left(\frac{4}{\pi}\frac{1}{(m\hbar\omega)^3}\right)^{\frac{1}{4}} e^{-\frac{i\pi}{2}} e^{-\frac{p^2}{2m\hbar\omega}}.$$

The phase factor can be omitted.

Solution: The normalized wave function for the ground state and the first excited states are

$$u_0(x) = \left(\frac{2\omega}{\pi\hbar}\right)^{1/4} e^{-(\frac{m\omega}{2\hbar})x^2}$$

$$u_1(x) = \left[\frac{4}{\pi}\left(\frac{m\omega}{\hbar}\right)^3\right]^{1/4} xe^{-(\frac{m\omega}{2\hbar})x^2}$$

Thus

$$\phi_0(p) = \frac{1}{\sqrt{2\pi\hbar}} \int e^{-ipx/\hbar} u_0(x) dx$$

$$= \frac{1}{\sqrt{2\pi\hbar}} \left(\frac{2\omega}{\pi\hbar}\right)^{1/4} \int e^{-ipx/\hbar} e^{-\frac{m\omega}{2\hbar}x^2} dx$$

$$= \frac{1}{\sqrt{2\pi\hbar}} \left(\frac{2\omega}{\pi\hbar}\right)^{1/4} e^{-\frac{p^2}{2m\hbar\omega}} \int e^{-\frac{m\omega}{2\hbar}z^2} dz$$

$$\phi_1(p) = \frac{1}{\sqrt{2\pi\hbar}} \left[\frac{4}{\pi}\left(\frac{m\omega}{\hbar}\right)^3\right]^{1/4} e^{-\frac{p^2}{2m\hbar\omega}}$$

$$\times \int \left(z - \frac{ip}{m\omega}\right) e^{-\frac{m\omega}{2\hbar}z^2} dz$$

where

$$z = x + \frac{ip}{m\omega}$$

Now

$$\int e^{-\frac{m\omega}{2\hbar}z^2} dz = \left(\frac{2\hbar}{m\omega}\right)^{1/2} \sqrt{\pi}$$

$$\int \left(z - \frac{ip}{m\omega}\right) e^{-\frac{m\omega}{2\hbar}z^2} dz = -ip \left(\frac{2\hbar}{(m\omega)^3}\right)^{1/2} \sqrt{\pi}$$

Hence

$$\phi_0(p) = \left(\frac{1}{\pi\hbar\omega}\right)^{1/4} e^{-\frac{p^2}{2m\hbar\omega}}$$

$$\phi_1(p) = \frac{1}{\sqrt{2\pi\hbar}} \left[\frac{4}{\pi}\left(\frac{m\omega}{\hbar}\right)^3\right]^{1/4} e^{-i\pi/2}$$

$$\times \left(\frac{2\hbar}{(m\omega)^3}\right)^{1/2} \sqrt{\pi} p e^{-\frac{p^2}{2m\hbar\omega}}$$

$$= \left[\frac{4}{\pi}\left(\frac{1}{m\omega\hbar}\right)^3\right]^{1/4} p e^{-\frac{p^2}{2m\hbar\omega}}$$

Chapter 6

Angular Momentum

Q6.1 Consider an operator V_+ which satisfies the commutation relations

$$[L_+, V_+] = 0$$
$$[L_z, V_+] = \hbar V_+.$$

Using these relations, show that

$$L_z(V_+ Y_{ll}) = (l+1)\hbar(V_+ Y_{ll})$$
$$L^2(V_+ Y_{ll}) = (l+1)(l+2)\hbar^2(V_+ Y_{ll}).$$

Solution: Using the commutation relation given in the problem

$$L_z(V_+ Y_{ll}) = ([L_z, V_+] + V_+ L_z)Y_{ll} = (l+1)\hbar(V_+ Y_{ll}) \quad (6.1)$$
$$[L^2, V_+] = [L_-, V_+]L_+ + L_-[L_+, V_+] + \hbar[L_z, V_+]$$
$$+ L_z[L_z, V_+] + [L_z, V_+]L_z$$

Using the commutation relations given in the problem, $L_+ Y_{ll} = 0$, we get

$$L^2(V_+ Y_{ll}) = (V_+ L^2 + 2\hbar V_+ L_z + 2\hbar^2 V_+)Y_{ll}$$
$$= [l(l+1) + 2l\hbar^2 + 2\hbar^2]V_+ Y_{ll}$$

Hence

$$L^2(V_+ Y_{ll}) = \hbar^2(l+1)(l+2)V_+ Y_{ll} \quad (6.2)$$

Q6.2 Show that when the system is described by a state function $\psi(\mathbf{r})$,

$$\Delta L_x \Delta L_y \geq \frac{\hbar}{2}|\langle L_z \rangle|$$

where

$$(\Delta L_x)^2 = \langle (L_x - \langle L_x \rangle)^2 \rangle$$
$$(\Delta L_y)^2 = \langle (L_y - \langle L_y \rangle)^2 \rangle$$

and $\langle L_x \rangle$ and $\langle L_y \rangle$ denote the average values of L_x and L_y in the state $\psi(\mathbf{r})$. If $\psi(\mathbf{r})$ is normalised eigenfunction of L_z, with eigenvalue $m\hbar$, what is $\Delta L_x \, \Delta L_y$?

Solution: Using the result given in Ch. 3 [c.f. Eq. (3.63)]

$$(\Delta L_x)^2 (\Delta L_y)^2 \geq \frac{1}{4}|\langle \psi|[L_x, L_y]|\psi \rangle|^2$$

$$= \frac{\hbar^2}{4}|\langle L_z \rangle|^2$$

Hence

$$(\Delta L_x)(\Delta L_y) \geq \frac{\hbar}{2}|\langle L_z \rangle| \qquad (6.3)$$

If $\psi(r)$ is a normalized eigenfunction of L_z, then

$$L_z \psi(r) = m\hbar \psi(r) \qquad (6.4)$$

$$\therefore (\Delta L_x)(\Delta L_y) \geq \left| \frac{m\hbar^2}{2} \right| \qquad (6.5)$$

Q6.3 Find the commutator

$$[L^2, x_i], \quad i = 1, 2, 3$$

by using the result

$$[x_i, \hat{p}_j] = i\hbar \delta_{ij}$$

Solution: Now

$$[L^2, x_i] = [L_j L_j, x_i]$$

$$L_j = \epsilon_{jkn} x_k p_n$$

$$\therefore [L_j, x_i] = \epsilon_{jkn}[x_k p_n, x_i]$$

$$= \epsilon_{jkn} x_k(-i\hbar\delta_{in}) = -i\hbar\epsilon_{ijk}x_k$$

Hence

$$[L^2, x_i] = L_j[L_j, x_i] + [L_j, x_i]L_j = -i\hbar\epsilon_{ijk}\{L_j x_k + x_k L_j\}$$

$$(6.6)$$

Q6.4 Show that for a state ψ_{lm} such that

$$L_z\psi_{lm} = m\hbar\psi_{lm}$$

$$L^2\psi_{lm} = l(l+1)\hbar^2\psi_{lm}$$

the average values of L_x^2, L_y^2 are given by

$$\langle L_x^2 \rangle = \langle L_y^2 \rangle = \frac{l(l+1)\hbar^2 - m^2\hbar^2}{2}$$

Solution: To find $\langle L_x^2 \rangle$ and $\langle L_y^2 \rangle$ we use the following relations

$$L_+ = L_x + iL_y \qquad (6.7)$$

$$L_- = L_x - iL_y \qquad (6.8)$$

Using Eqs. (6.7) and (6.8) we get

$$L_x^2 = \frac{1}{4}(L_+^2 + L_+L_- + L_-L_+ + L_-^2) \qquad (6.9)$$

$$L_y^2 = -\frac{1}{4}(L_+^2 - L_+L_- - L_-L_+ + L_-^2) \qquad (6.10)$$

Now

$$L_+\psi_{lm} = \hbar\sqrt{(l-m)(l+m+1)}\psi_{lm+1} \qquad (6.11)$$

$$L_-\psi_{lm} = \hbar\sqrt{(l+m)(l-m+1)}\psi_{lm-1} \qquad (6.12)$$

Using Eqs. (6.11) and (6.12), we get from Eq. (6.9) [note that only L_+L_-, or L_-L_+ contribute]

$$\langle L_x^2 \rangle = \frac{\hbar^2}{4}[(l - m + 1)(l + m) + (l + m + 1)(l - m)]$$

or

$$\langle L_x^2 \rangle = \frac{l(l + 1)\hbar^2 - m^2\hbar^2}{2}$$

Similarly

$$\langle L_y^2 \rangle = \frac{l(l + 1)\hbar^2 - m^2\hbar^2}{2}$$

Q6.5 Show that for an eigenstate ψ_{lm} of L_z, the average value of L_x is zero.

Solution: The average value of L_x:

$$\langle L_x \rangle = \frac{1}{2}\langle \psi_{lm}|(L_+ + L_-)|\psi_{lm}\rangle \qquad (6.13)$$

Using Eqs. (6.11), (6.12), (6.13) and orthogonality of the states involved, one gets

$$\langle L_x \rangle = 0$$

Q6.6 Show that

$$[L^2, [L^2, \mathbf{r}]] = 2\hbar^2(\mathbf{r}L^2 + L^2\hat{\mathbf{r}}).$$

Solution: To find the commutation relation $[L^2, [L^2, \mathbf{r}]]$ we use Eq. (6.6) and $[L^2, L_i] = 0$.
Thus

$$
\begin{aligned}
[L^2, [L^2, x_k]] &= -i\hbar\epsilon_{ikl}\{[L^2, L_i x_l] + [L^2, x_l L_i]\} \\
&= i\hbar\epsilon_{kil}\{L_i[L^2, x_l] + [L^2, x_l]L_i\} \\
&= (i\hbar)^2\epsilon_{kil}\epsilon_{rlm}\{L_i(L_r x_n + x_n L_r) \\
&\quad + (L_r x_n + x_n L_r)L_i\} \qquad (6.14)
\end{aligned}
$$

Using the following relations

$$\epsilon_{lik}\epsilon_{lrn} = (\delta_{ir}\delta_{kn} - \delta_{in}\delta_{kr})$$

and

$$x_k L_i = L_i x_k + i\hbar\epsilon_{ikr} x_r$$

$$L_k x_i = -x_i L_k - i\hbar\epsilon_{kir} x_r$$

we obtain from Eq. (6.14):

$$[L^2, [L^2, x_k]] = \hbar^2 \{2L^2 x_k + 2x_k L^2 - L_k(\mathbf{L}\cdot\mathbf{r} + \mathbf{r}\cdot\mathbf{L})$$

$$- (\mathbf{L}\cdot\mathbf{r} + \mathbf{r}\cdot\mathbf{L})L_k\}$$

Now

$$\mathbf{L}\cdot\mathbf{r} + \mathbf{r}\cdot\mathbf{L} = 0$$

Hence we have

$$[L^2, [L^2, x_k]] = 2\hbar^2(x_k L^2 + L^2 x_k)$$

or

$$[L^2, [L^2, \mathbf{r}] = 2\hbar^2(\mathbf{r}L^2 + L^2\mathbf{r})$$

Q6.7 Using the relation derived in the text

$$L_\pm Y_{lm}(\theta, \phi) = \sqrt{(l \mp m)(l \pm m + 1)}\hbar Y_{l,m\pm 1}(\theta, \phi)$$

and the expressions for $L_\pm$ in spherical polar coordinates, show that

$$P_l^{m\pm 1}(\theta) = [(l \mp m)(l \pm m + 1)]^{-1/2}$$

$$\times \left(\pm\frac{\partial}{\partial\theta} - m\cot\theta\right) P_l^m(\theta)$$

$$= [(l \mp m)(l \pm m + 1)]^{-1/2}(\pm)\sin^{\pm m}\theta\frac{d}{d\theta}$$

$$\times (\sin^{\mp m}\theta P_{lm}(\theta)). \tag{6.15}$$

Solution: In spherical polar coordinates

$$L_\pm = \hbar e^{\pm i\phi}\left(\pm\frac{\partial}{\partial\theta} + i\cot\theta\frac{\partial}{\partial\phi}\right)$$

Thus

$$\hbar e^{\pm i\phi}\left(\pm\frac{\partial}{\partial\theta}+i\cot\theta\frac{\partial}{\partial\phi}\right)P_l^m(\theta)\Phi_m(\phi)$$

$$= L_\pm Y_{lm}(\theta,\phi) = \sqrt{(l\mp m)(l\pm m+1)}Y_{l,m_\pm}(\theta,\phi)$$

$$= \sqrt{(l\mp m)(l\pm m+1)}P_l^{m\pm1}(\theta)\Phi_{m\pm1}(\phi)$$

where we have used the relation

$$Y_{l,m_\pm}(\theta,\phi) = P_l^m(\theta)\Phi_m(\phi)$$

Thus

$$P_l^{m\pm1}(\theta)\Phi_{m\pm1}(\phi)$$

$$= [(l\mp m)(l\pm m+1)]^{-1/2}e^{\pm i\phi}$$

$$\times\left(\pm\frac{\partial}{\partial\theta}+i\cot\theta\frac{\partial}{\partial\phi}\right)P_l^m(\theta)\Phi_m(\phi)$$

$$= [(l\mp m)(l\pm m+1)]^{-1/2}e^{\pm i\phi}$$

$$\times\left(\pm\frac{\partial}{\partial\theta}-m\cot\theta\right)P_l^m(\theta)\Phi_m(\phi)$$

$$= [(l\mp m)(l\pm m+1)]^{-1/2}$$

$$\times\left(\pm\frac{\partial}{\partial\theta}-m\cot\theta\right)P_l^m(\theta)e^{\pm i\phi}\Phi_m(\phi)$$

Now

$$L_\pm Y_{lm}(\theta,\phi) = \sqrt{(l\mp m)(l\pm m+1)}\hbar Y_{l,m\pm1}(\theta,\phi) \tag{6.16}$$

Using Eqs. (6.11a) and (6.11b) from the text, we have from Eq. (6.16)

$$L_\pm Y_{lm}(\theta,\phi) \equiv \hbar e^{\pm i\phi}\left(\pm\frac{\partial}{\partial\theta}+i\cot\theta\frac{\partial}{\partial\phi}\right)(2\pi)^{-\frac{1}{2}}e^{im\phi}\Theta_{lm}(\theta)$$

$$= \sqrt{(l\mp m)(l\pm m+1)}\hbar\Phi_{m\pm1}(\theta)\Theta_{lm\pm1}(\theta) \tag{6.17}$$

Therefore

$$\hbar\Phi_{m\pm1}(\phi)\left(\pm\frac{\partial}{\partial\theta} - m\cot\theta\right)\Theta_{lm}(\theta)$$

$$= \sqrt{(l\mp m)(l\pm m+1)}\hbar\Phi_{m\pm1}(\theta)\Theta_{lm\pm1}(\theta) \quad (6.18)$$

Hence

$$\Theta_{l,m\pm1}(\theta) = [(l\mp m)(l\pm m+1)]^{-1/2}$$

$$\times \left(\pm\frac{\partial}{\partial\theta} - m\cot\theta\right)\Theta_{lm}(\theta)$$

$$= [(l\mp m)(l\pm m+1)]^{-1/2}(\pm)\sin^{\pm m}\theta$$

$$\times \frac{d}{d\theta}(\sin^{\mp m}\theta\Theta_{lm}(\theta)) \quad (6.19)$$

Putting $m = -l$ in Eq. (6.19) and using lower sign, we must have

$$\frac{d}{d\theta}(\sin^{-l}\theta\Theta_{l-l}(\theta)) = 0 \quad (6.20)$$

This is a consequence of the fact that there is no state with

$$m < -l \quad (6.21)$$

Thus we have

$$\sin^{-l}\theta\Theta_{l-l}(\theta) = constant = a \quad (6.22)$$

or

$$\Theta_{l-l}(\theta) = a\sin^l\theta \quad (6.23)$$

The constant a can be determined from the normalisation Eq. (6.21b) of the text

$$a^2\int_0^\pi(\sin\theta)^{2l+1}d\theta = 1 \quad (6.24)$$

so that

$$a^2 \times \frac{2.2^{2l}(l!)^2}{(2l+1)!} = 1 \quad (6.25)$$

or

$$a = \frac{1}{2^l l!} \left(\frac{(2l+1)!}{2} \right)^{1/2} \tag{6.26}$$

Now from Eq. (6.17)

$$\Theta_{lm+1}(\theta) = ((l-m)(l+m+1))^{-1/2}(-1)\sin^{m+1}\theta \frac{d}{d(\cos\theta)}$$

$$\times (\sin^{-m}\theta\Theta_{lm}(\theta)). \tag{6.27}$$

Thus changing $m \to m-1$, we have

$$\Theta_{lm}(\theta) = ((l-m+1)(l+m))^{-1/2}(-1)\sin^m\theta \frac{d}{d(\cos\theta)}$$

$$\times (\sin^{-m+1}\theta\Theta_{lm-1}(\theta)). \tag{6.28}$$

Using (6.28) twice, we obtain

$$\Theta_{lm}(\theta) = ((l-m+1)(l+m))^{-1/2}(-1)\sin^m\theta \frac{d}{d(\cos\theta)}$$

$$\times (\sin^{-m+1}\theta\{(l-m+2)(l+m-1)\}^{-1/2}$$

$$\times (-1)\sin^{m-1}\theta \frac{d}{d(\cos\theta)}(\sin^{-m+2}\theta\Theta_{lm-2}(\theta)))$$

$$= (-1)^2((l-m+1)(l-m+2)$$

$$\times (l+m)(l+m-1))^{-1/2}$$

$$\times \sin^m\theta \frac{d^2}{d(\cos\theta)^2}(\sin^{-m+2}\theta\Theta_{lm-2}(\theta)).$$

Repeated application gives

$$\Theta_{lm}(\theta) = (-1)^{l+m}((l-m+1)(l-m+2)\cdots$$

$$\times 2l(l+m)(l+m-1)\cdots 1)^{-1/2}$$

$$\times \sin^m\theta \frac{d^{l+m}}{d(\cos\theta)^{l+m}}\left(\sin^l\theta\Theta_{l-l}(\theta) \right).$$

Therefore

$$\Theta_{lm}(\theta)$$

$$= (-1)^{l+m} \left((l+m)! \frac{2l \cdots (l-m+1)(l-m) \cdots 1}{(l-m) \cdots 1} \right)^{-1/2}$$

$$\times \sin^m \theta \left(\frac{d}{d(\cos \theta)} \right)^{l+m} (\sin^l \theta a \sin^l \theta)$$

$$= (-1)^{l+m} a \left(\frac{(l+m)! 2l!}{(l-m)!} \right)^{-1/2} \sin^m \theta \left(\frac{d}{d(\cos \theta)} \right)^{l+m}$$

$$\times (\sin \theta)^{2l} \tag{6.29}$$

Putting the value of a from Eq. (6.28), we get

$$\Theta_{lm}(\theta = (-1)^{l+m} \frac{1}{2^l l!} \left(\frac{(2l+1)(l-m)!}{2(l+m)!} \right)^{1/2}$$

$$\times \sin^m \theta \left(\frac{d}{d(\cos \theta)} \right)^{l+m} (\sin \theta)^{2l} \tag{6.30}$$

Hence the normalised simultaneous eigenfunctions of L^2 and L_z are given by

$$Y_{lm}(\theta, \phi) = \Theta_{lm}(\theta) \Phi_m(\phi)$$

$$= (-1)^{l+m} \frac{1}{2^l l!} \left(\frac{(2l+1)(l-m)!}{4\pi(l+m)!} \right)^{1/2}$$

$$\times \sin^m \theta \left(\frac{d}{d(\cos \theta)} \right)^{l+m} (\sin \theta)^{2l} e^{im\phi}$$

$$= (-1)^m \left(\frac{(2l+1)(l-m)!}{4\pi(l+m)!} \right)^{1/2} P_l^m(\cos \theta) e^{im\phi} \tag{6.31}$$

$Y_{lm}(\theta, \phi)$ are called normalised spherical harmonics. In (6.31) P_l^m are given by

$$P_l^m(\cos \theta) = \frac{(-1)^l}{2^l l!} \sin^m \theta \left(\frac{d}{d(\cos \theta)} \right)^{l+m} (\sin \theta)^{2l}$$

or

$$P_l^m(\omega) = \frac{(1-\omega^2)^{m/2}}{2^l l!} \frac{d^{l+m}}{d\omega^{l+m}} (\omega^2 - 1)^l,$$

where $\omega = \cos\theta$, $P_l^m(\omega)$ are known as Associated Legendre Polynomials.

Chapter 7

Motion in Centrally Symmetric Field

Q7.1 In spectroscopic notation, a state is specified as $^{2s+1}l_j$, where s is the spin and j is the total angular momentum. In particular for hydrogen atom, $s = \frac{1}{2}$ i.e the spin of the electron. Thus $j = l - \frac{1}{2}, l + \frac{1}{2}$. List all the possible states for $n = 1, n = 2, n = 3$. Find $\left(\frac{1}{\lambda_3} - \frac{1}{\lambda_2}\right), \left(\frac{1}{\lambda_3} - \frac{1}{\lambda_1}\right), \left(\frac{1}{\lambda_2} - \frac{1}{\lambda_1}\right)$. Show that for the transitions from the level

$$n = 3 \quad \text{to} \quad n = 2,$$
$$n = 3 \quad \text{to} \quad n = 1,$$
$$n = 2 \quad \text{to} \quad n = 1,$$

the photons of wavelengths $6559\text{Å}, 1025\text{Å}$ and 1215Å are emitted respectively.

Solution: We know $n \geq \ell + 1$, so

$$n = 1 \quad \ell = 0$$
$$n = 2 \quad \ell = 0, 1$$
$$n = 3 \quad \ell = 0, 1, 2$$

and

$$\ell = 0 \quad j = 1/2$$
$$\ell = 1 \quad j = 1/2, 3/2$$
$$\ell = 2 \quad j = 3/2, 5/2$$

Thus all possible states are

$$n = 1 \Rightarrow {}^2 s_{1/2}$$
$$n = 2 \Rightarrow {}^2 s_{1/2}, {}^2 p_{1/2}, {}^2 p_{3/2}$$
$$n = 3 \Rightarrow {}^2 s_{1/2}, {}^2 p_{1/2}, {}^2 p_{3/2}, {}^2 d_{3/2}, {}^2 d_{5/2}$$

Now

$$E_n = -Z^2 \frac{(2\pi \hbar c) R}{n^2}$$

where $Z = 1$ for the hydrogen atom and the Rydberg constant is

$$R = \frac{\alpha^2 (m_e c^2)}{4\pi \hbar c} \simeq 1.0974 \times 10^5 \, \text{cm}^{-1}$$

Now

$$E_n = 2\pi \hbar \nu = \frac{2\pi \hbar c}{\lambda_n}$$

Hence

$$\frac{1}{\lambda_n} = -\frac{R}{n}$$

Now

$$\Delta \left(\frac{1}{\lambda} \right)_{32} = \frac{1}{\lambda_3} - \frac{1}{\lambda_2} = -R \left(\frac{1}{9} - \frac{1}{4} \right)$$

$$= \frac{5R}{36} = 1.5241 \times 10^8 \, \text{cm}^{-1}$$

$$\Delta \left(\frac{1}{\lambda} \right)_{31} = \frac{1}{\lambda_3} - \frac{1}{\lambda_1} = -R \left(\frac{1}{9} - 1 \right)$$

$$= \frac{8R}{9} = 9.7544 \times 10^8 \, \text{cm}^{-1}$$

$$\Delta \left(\frac{1}{\lambda} \right)_{21} = \frac{1}{\lambda_2} - \frac{1}{\lambda_1} = -R \left(\frac{1}{4} - 1 \right)$$

$$= \frac{3R}{4} = 8.2303 \times 10^8 \, \text{cm}^{-1}$$

Hence

$$n = 3 \quad \text{to} \quad n = 2, \quad \Delta\lambda_{3\to 2} \approx 6561\text{Å}$$
$$n = 3 \quad \text{to} \quad n = 1, \quad \Delta\lambda_{3\to 1} \approx 1025\text{Å}$$
$$n = 2 \quad \text{to} \quad n = 1, \quad \Delta\lambda_{2\to 1} \approx 1215\text{Å}$$

Q7.2 A particle of mass m moves in a square well potential

$$V(\mathbf{r}) = -V_0 \quad r < a$$
$$= 0 \qquad r > a.$$

If bound s-state exists, show that energy eigenvalues are given by

$$-\gamma = k \cot \, ka$$

where

$$\gamma = \sqrt{2mW/\hbar^2}, \quad E = -W$$
$$k = \sqrt{2m(V_0 - W)/\hbar^2}.$$

If there is only one such state and it is very weakly bound, show that

$$V_0 a^2 \sim \frac{\pi^2 \hbar^2}{8m}$$

Solution: Radial equation for $\ell = 0$ is

$$\left(-\frac{\hbar^2}{2\mu} \frac{d^2}{dr^2} - V_0 - E \right) \chi = 0 \quad r < a$$

$$\left(-\frac{\hbar^2}{2\mu} \frac{d^2}{dr^2} - E \right) \chi = 0 \quad r > a$$

Put

$$\gamma = \sqrt{-\frac{2\mu E}{\hbar^2}} = \sqrt{\frac{2\mu|E|}{\hbar^2}}$$

$$k = \sqrt{\frac{2\mu(V_0 + E)}{\hbar^2}} = \sqrt{\frac{2\mu(V_0 - |E|)}{\hbar^2}}$$

Now

$$\left(\frac{d^2}{dr^2} + k^2\right)\chi = 0 \quad r < a \tag{7.1}$$

$$\left(\frac{d^2}{dr^2} - \gamma^2\right)\chi = 0 \quad r > a \tag{7.2}$$

The solution of Eq. (7.1) satisfying the boundary condition:

$$\chi(0) = 0$$

is given by

$$\chi(r) = A \sin kr \quad r < a$$

Solution of Eq. (7.2) is

$$\chi(r) = Be^{-\gamma r} + Ce^{\gamma r}, \quad r > a$$

$e^{\gamma r}$ is not possible as $\chi(r) \to 0$ when $r \to \infty$, so

$$\chi(r) = Be^{-\gamma r}, \quad r > a$$

Boundary conditions at $r = a$ give

$$A \sin ka = Be^{-\gamma a}$$

$$Ak \cos ka = -B\gamma e^{-\gamma a}$$

From the above equations, we get

$$k \cot ka = -\gamma$$

i.e. the energy eigenvalues. For very weakly bound states,

$$|E| \ll V_0, \quad \frac{|E|}{V_0} \ll 1$$

$$\cot^2 ka = \frac{\gamma^2}{k^2} = \frac{|E|/V_0}{1 - |E|/V} \to 0$$

which gives

$$ka = \pi/2$$

or

$$k^2 a^2 = \frac{\pi^2}{4}$$

Hence

$$V_0 a^2 = \frac{\pi^2 \hbar^2}{8\mu}$$

Q7.3 Assume that interaction between the neutron and the proton that makes up a deuteron can be represented by a square well potential with $a \sim 2 \times 10^{-13}$ cm. If $l = 0$ energy level of the system (binding energy) is 2 MeV, calculate V_0 in MeV.

Solution: The binding energy of the deuteron, $E = -2$ MeV, is much smaller than the strong nuclear potential. Thus $\frac{|E|}{V_0} \ll 1$. Hence we can use the relation (derived in 7.2)

$$V_0 a^2 = \frac{\pi^2 \hbar^2}{8\mu}$$

For the deuteron, the reduced mass

$$\mu = \frac{m_p m_n}{m_p + m_n} \simeq \frac{1}{2} m_p, \quad m_p \approx m_n$$

For $a = 2 \times 10^{-13}$ cm

$$V_0 = \frac{\pi^2 \hbar^2}{8a^2 \mu} = \frac{\pi^2 \hbar^2 c^2}{4 m_p c^2 a^2}$$

$$= \frac{\pi^2 (197 \text{MeV} \times 10^{-13} \text{ cm})^2}{4(940) \text{MeV} (2 \times 10^{-13})^2 \text{ cm}^2}$$

$$= 25.5 \text{ MeV}$$

Thus $\frac{|E|}{V_0} \approx 0.08$, a good approximation for the deuteron.

Q7.4 The radial wave function $\chi(\mathbf{r})$ satisfies the equation

$$\frac{d^2 \chi}{dr^2} + \left(\frac{2m}{\hbar^2} (E - V) - \frac{l(l+1)}{r^2} \right) \chi = 0$$

Take $V(r) = (1/2) m \omega^2 r^2$ and by expanding $\chi \exp(1/2\rho^2)$ in powers of $\rho = r/b$ with $b^2 = \hbar/m\omega$, show that energy levels are given by

$$E_n = (1/2)\hbar\omega(4n + 2l + 3)$$

where n is an integer ≥ 0.

Solution: We have

$$V(r) = \frac{1}{2}\mu\omega^2 r^2$$

$$\chi(r) = rR(r)$$

$$\chi(0) = 0$$

Radial equation for energy eigenvalues is

$$\frac{d^2\chi}{dr^2} + \left(\frac{2\mu}{\hbar^2}(E - \frac{1}{2}\mu\omega^2 r^2) - \frac{\ell(\ell+1)}{r^2}\right)\chi(r) = 0 \quad (7.3)$$

Introduce the dimensionless variables

$$\rho = r/b = \sqrt{\frac{m\omega}{\hbar}}r, \quad \epsilon = \frac{2}{\hbar\omega}E$$

With these variables, Eq. (7.3) becomes

$$\frac{d^2\chi}{d\rho^2} + \left(\epsilon - \rho^2 - \frac{\ell(\ell+1)}{\rho^2}\chi(\rho)\right) = 0 \quad (7.4)$$

For $\rho \to 0$, $\frac{1}{\rho^2}$ dominates and one has the solution

$$\chi(\rho) \sim \rho^{\ell+1}$$

as in the hydrogen atom. But for $\rho \to \infty$,

$$\frac{d^2}{d\rho^2} - \rho^2\chi = 0$$

which has solution in this limit

$$\chi(\rho) \sim e^{-\frac{1}{2}\rho^2}$$

Thus we can write

$$\chi(\rho) = \rho^{\ell+1}e^{-\frac{1}{2}\rho^2}K(\rho)$$

Substituting in Eq. (7.4), we get

$$\frac{d^2 K(\rho)}{d\rho^2} + [2(\ell+1) - 2\rho^2]\frac{dK(\rho)}{d\rho} - [2\ell + 3 - \epsilon]K(\rho) = 0$$

$$(7.5)$$

Writing the solution of the above equation as power series

$$K(\rho) = \sum a_\nu \rho^\nu$$

and substituting in Eq. (7.5) we get,

$$\sum_\nu [\{a_\nu \nu(\nu - 1 + 2(l+1))\}\rho^{\nu-2}$$

$$-a_\nu \{2\nu + (2l + 3 - \epsilon)\}\rho^\nu] = 0$$

This gives on equating the coefficient of $\rho^\nu = 0$

$$[(\nu+2)((\nu+1) + 2(\ell+1))]a_{\nu+2}$$

$$-[2\nu + 2\ell + 3 - \epsilon]a_\nu = 0$$

or

$$[(\nu+1)(\nu+2) + 2(\ell+1)(\nu+2)]a_{\nu+2}$$

$$= (2\ell + 3 + 2\nu - \epsilon)a_\nu$$

$$(7.6)$$

For $\nu = -1$

$$a_1 = 0$$

and for $\nu = 1$

$$a_3 = 0$$

Thus

$$a_\nu = 0 \quad \text{for odd } \nu$$

Put

$$\nu = 2n \quad n = 0, 1, \ldots$$

Then from Eq. (7.6)

$$\frac{a_{2n+2}}{a_{2n}} = \frac{(2\ell + 3 + 4n - \epsilon)}{[(2n+1)(2n+2) + 2(\ell+1)(2n+2)]} \quad (7.7)$$

Compare with the known series

$$e^{\rho^2} = 1 + \rho^2 + \frac{1}{2!}\rho^4 + \cdots$$

whose ratio of the coefficients of successive power gives

$$\frac{a_{n+1}}{a_n} = \frac{n!}{(n+1!)} = \frac{1}{n+1} \to \frac{1}{n}, \qquad \text{for large } n$$

Thus for large n

$$K(\rho) \sim e^{\rho^2}$$

so

$$\chi(\rho) \sim e^{\rho^2} e^{-\frac{1}{2}\rho^2}$$

$$= e^{\frac{1}{2}\rho^2} \quad \text{for large } \rho$$

which is not allowed, Hence the power series must terminate and $K(\rho)$ must be a polynomial. To ensure this

$$a_{2n+2} = 0$$

Hence from Eq. (7.7), since $a_{2n} \neq 0$, we get

$$(2\ell + 3 + 4n - \epsilon) = 0$$

or

$$\epsilon_{\ell n} = (4n + 2\ell + 3)$$

Hence using the relation

$$\epsilon_{n\ell} = \frac{2}{\hbar\omega} E_{nl}$$

so

$$E_{n\ell} = \frac{\hbar\omega}{2}(4n + 2\ell + 3)$$

where n is an integer ≥ 0. Note that $E_{n\ell}$ does depend here on ℓ, [in contrast to hydrogen atom] i.e. the orbital angular momentum in addition to principal quantum number n. To see the significance of above result, we see the energy spectrum is given by

$$E_N = \frac{1}{2}\hbar\omega(2N + 3)$$

which depends on the quantum number $N = 2n + \ell$. Thus all states with $\ell = N, N-2, \cdots, 0$ or 1 are degenerate i.e. have the same energy.

Some of the lower energy levels are listed below:

N	ℓ	n	state	parity	eigenvalue
0	0	0	$1s$	$+$	$\frac{3}{2}\hbar\omega$
1	1	0	$1p$	$-$	$\frac{5}{2}\hbar\omega$
2	$\begin{Bmatrix}2\\0\end{Bmatrix}$	$\begin{Bmatrix}0\\1\end{Bmatrix}$	$\begin{Bmatrix}1d\\2s\end{Bmatrix}$	$+$	$\frac{7}{2}\hbar\omega$
3	$\begin{Bmatrix}3\\1\end{Bmatrix}$	$\begin{Bmatrix}0\\1\end{Bmatrix}$	$\begin{Bmatrix}1f\\2p\end{Bmatrix}$	$-$	$\frac{9}{2}\hbar\omega$
4	$\begin{Bmatrix}4\\2\\0\end{Bmatrix}$	$\begin{Bmatrix}0\\1\\2\end{Bmatrix}$	$\begin{Bmatrix}1g\\2d\\3s\end{Bmatrix}$	$+$	$\frac{11}{2}\hbar\omega$

Thus energy levels are degenerate. For each ℓ, there are $(2\ell + 1)$ eigenvalues of m. Therefore degree of degeneracy

$$D_n = \sum (2\ell + 1) = \sum_{n=0}^{N/2}(2N - 4n + 1)$$

$$= \frac{1}{2}(N + 1)(N + 2)$$

Q7.5 A particle moves in a spherically symmetrical potential $V = \alpha/r^2 - \beta/r$, where α and β are constants. Find bound state energy levels.

Solution: Radial equation for energy eigenvalue for the potential

$$V(r) = -\frac{\beta}{r} + \frac{\alpha}{r^2}$$

is

$$\frac{d^2\chi}{dr^2} + \frac{2m}{\hbar^2}\left[E + \frac{\beta}{r} - \frac{\alpha}{r^2} - \frac{\ell(\ell+1)\hbar^2}{2mr^2}\right]\chi = 0 \quad (7.8)$$

Changing the variable to $y = \frac{r}{a}$, Eq. (7.8) can be written in the form

$$\frac{d^2\chi}{dy^2} + \left[-\epsilon^2 + \frac{K}{y} - \frac{\ell(\ell+1) + v_0}{y^2}\right]\chi = 0 \qquad (7.9)$$

where

$$\epsilon^2 = -\frac{2ma^2}{\hbar^2}E, \quad K = \frac{2ma}{\hbar^2}\beta, \quad v_0 = \frac{2m\alpha}{\hbar^2}$$

Put

$$\ell(\ell+1) + v_0 = b(b+1) \qquad (7.10)$$

Then Eq. (7.9) becomes

$$\frac{d^2\chi}{dy^2} + \left[-\epsilon^2 + \frac{K}{y} - \frac{b(b+1)}{y^2}\right]\chi = 0 \qquad (7.11)$$

The above equation is similar to the equation

$$\frac{d^2\chi}{dy^2} + \left[-\epsilon^2 + \frac{K}{y} - \frac{\ell(\ell+1)}{y^2}\right]\chi = 0 \qquad (7.12)$$

for the Coulomb potential of the H-atom given in the text with $K = \sqrt{\frac{2m}{\hbar^2}e^2}$, for which

$$E = -\epsilon^2 = -\frac{K^2}{4(n' + \ell + 1)}$$

where n' is an integer ≥ 0. Thus the energy eigenvalues for the problem (7.5) are given by (replace ℓ by b, $K = \frac{2ma}{\hbar^2}\beta$

and E by $\frac{2ma^2}{\hbar^2}E$)

$$E = -\frac{m\beta^2}{2\hbar^2}\frac{1}{[n'+1+b]^2}$$

$$= -\frac{m\beta^2}{2\hbar}\frac{1}{[(n'+1/2)+(b+1/2)]^2}$$

where

$$b(b+1) = \ell(\ell+1) + v_0$$

or

$$(b+1/2) = \left[1 + 4[\ell(\ell+1) + v_0]\right]^{1/2}/2$$

Note that we have selected the positive root.

Q7.6 Find the expectation values $\langle r \rangle$ and $\langle \frac{1}{r} \rangle$ for the ground state of hydrogen like atom.

Solution: The ground state wave function is

$$u_{100} = R_{10}Y_{00}$$

$$= 2\left(\frac{Z}{a_0}\right)^{3/2} e^{-Zr/a_0}Y_{00}$$

so

$$\langle r \rangle_{10} = \int_0^\infty u_{100}^*(r)r u_{100}(r)d^3r$$

$$= 4\left(\frac{Z}{a_0}\right)^3 \int_0^\infty e^{-2Zr/a_0}r^3 dr \underbrace{\int Y_{00}^* Y_{00}d\Omega}_{=1}$$

Using the integral

$$\int_0^\infty r^n e^{-br}dr = \frac{n!}{b^{n+1}}$$

we get

$$\langle r \rangle_{10} = 4 \left(\frac{Z}{a_0} \right)^3 \frac{3!}{(2Z/a_0)^4}$$

$$= \frac{3}{2} \frac{a_0}{Z}$$

Now

$$\left\langle \frac{1}{r} \right\rangle_{10} = 4 \left(\frac{Z}{a_0} \right)^3 \int_0^\infty r e^{-2Zr/a_0} dr$$

$$= 4 \left(\frac{Z}{a_0} \right)^3 \left(\frac{a_0}{2Z} \right)^2 = \frac{Z}{a_0}$$

Q7.7 Find the energy eigenvalues and eigenfunctions for a three-dimensional harmonic oscillator. The potential for such an oscillator is given by

$$V(x, y, z) = \frac{m}{2} (\omega_1^2 x^2 + \omega_2^2 y^2 + \omega_3^2 z^2)$$

Discuss the degeneracy of energy levels. Hint: Write the wave function as

$$u(x, y, z) = u_1(x) u_2(y) u_3(z)$$

and show that each of the functions $u_1(x)$, $u_2(y)$ and $u_3(z)$ satisfy one dimensional harmonic oscillator equation.

Solution: The Schrödinger equation in the cartesian co-ordinate system for the potential $V(x, y, z)$ is given by

$$\left[-\frac{\hbar^2}{2m} \left(\frac{\partial^2}{\partial x^2} + \frac{\partial^2}{\partial y^2} + \frac{\partial^2}{\partial z^2} \right) + \frac{m}{2} (\omega_1^2 x^2 + \omega_2^2 y^2 + \omega_3^2 z^2) \right]$$
$$\times u_1(x) u_2(y) u_3(z) = 0$$

where the nature of the problem suggests that we try $\psi(x, y, z) = u_1(x)u_2(y)u_3(z)$. Thus

$$u_2(y)u_3(z)\left[-\frac{\hbar^2}{2m}\frac{\partial^2 u_1(x)}{\partial x^2}\right]$$

$$+u_1(x)u_3(z)\left[-\frac{\hbar^2}{2m}\frac{\partial^2 u_2(y)}{\partial y^2}\right]$$

$$+u_1(x)u_2(y)\left[-\frac{\hbar^2}{2m}\frac{\partial^2 u_3(z)}{\partial z^2}\right]$$

$$+\frac{m}{2}(\omega_1^2 x^2 + \omega_2^2 y^2 + \omega_3^2 z^2)u_1(x)u_2(y)u_3(z)$$

$$= Eu_1(x)u_2(y)u_3(z)$$

or

$$\left[\frac{1}{u_1(x)}\left(-\frac{\hbar^2}{2m}\frac{\partial^2 u_1(x)}{\partial x^2}\right) + \frac{m}{2}\omega_1^2 x^2\right]$$

$$+\left[\frac{1}{u_2(x)}\left(-\frac{\hbar^2}{2m}\frac{\partial^2 u_2(y)}{\partial y^2}\right) + \frac{m}{2}\omega_2^2 y^2\right]$$

$$+\left[\frac{1}{u_3(z)}\left(-\frac{\hbar^2}{2m}\frac{\partial^2 u_3(z)}{\partial z^2}\right) + \frac{m}{2}\omega_3^2 z^2\right] = E$$

Now the three terms in the square brackets are respectively a function of x only, y only and z only while the right hand is a constant. Thus the only way we satisfy the equation is to have

$$\frac{1}{u_1(x)}\left[-\frac{\hbar^2}{2m}\frac{\partial^2 u_1(x)}{\partial x^2}\right] + \frac{m}{2}\omega_1^2 x^2 = E_1$$

$$\frac{1}{u_2(y)}\left[-\frac{\hbar^2}{2m}\frac{\partial^2 u_2(y)}{\partial y^2}\right] + \frac{m}{2}\omega_2^2 y^2 = E_2$$

$$\frac{1}{u_3(z)}\left[-\frac{\hbar^2}{2m}\frac{\partial^2 u_3(z)}{\partial z^2}\right] + \frac{m}{2}\omega_1^2 z^2 = E_3$$

where $E_1 + E_2 + E_3 = E$

Hence we have three differential equations for three one dimensional oscillators:

$$\frac{d^2u_1(x)}{dx^2} + \frac{2m}{\hbar^2}\left(E_1 - \frac{1}{2}m\omega_1^2 x^2\right)u_1(x) = 0$$

$$\frac{d^2u_2(y)}{dy^2} + \frac{2m}{\hbar^2}\left(E_2 - \frac{1}{2}m\omega_2^2 y^2\right)u_2(y) = 0$$

$$\frac{d^2u_3(z)}{dz^2} + \frac{2m}{\hbar^2}\left(E_3 - \frac{1}{2}m\omega_3^2 z^2\right)u_3(z) = 0$$

Hence the energy eigenvalues are given by

$$E = E_1 + E_2 + E_3 = \left[\left(n_1 + \frac{1}{2}\right)\hbar\omega_1\right.$$

$$+ \left.\left(n_2 + \frac{1}{2}\right)\hbar\omega_2 + \left(n_3 + \frac{1}{2}\right)\hbar\omega_3\right]$$

Q7.8 For

$$H = \frac{\hat{p}^2}{2m} + V(\mathbf{r}) = \hat{T} - \frac{Ze^2}{\mathbf{r}}$$

Show that

$$\langle \hat{T}\rangle_{nl} = -\frac{1}{2}\langle V\rangle_{nl}.$$

Solution:

$$E_n = \langle T\rangle_{n\ell} + \langle V\rangle_{n\ell}$$

and

$$\langle V\rangle_{nl} = -Ze^2 \left\langle\frac{1}{r}\right\rangle_{nl}$$

Using

$$\left\langle\frac{1}{r}\right\rangle_{nl} = \frac{Z}{a_0 n^2},$$

$$\langle V\rangle_{nl} = -\frac{Z^2 e^2}{a_0 n^2}$$

and

$$E_n = -\frac{Z^2 e^2}{2a_0 n^2}$$

we get

$$\langle T \rangle_{n\ell} = E_n - \langle V \rangle_{n\ell}$$

$$= \frac{1}{2}\frac{Z^2 e^2}{2a_0 n^2}$$

$$= -\frac{1}{2}\langle V \rangle_{n\ell}$$

Chapter 8

Collision Theory

Q8.1 The wave function

$$u(r, \theta, \phi) \xrightarrow[r \to \infty]{} A \left(e^{ikz} + f(\theta, \phi) \frac{e^{ikr}}{r} \right)$$

represents the incident plane wave and the scattered wave radially moving outward. Let **S** denotes the probability current density. Show that

$$\mathbf{S}_{\text{out}} = \frac{v}{r^3} |A|^2 |f(\theta, \phi)|^2 \mathbf{r}$$

$\quad +$ higher terms which can be neglected for large r.

Calculate $\mathbf{S}_{\text{out}} \cdot d\mathbf{\Sigma}$ where $|d\mathbf{\Sigma}| = d\Sigma$ is the surface element $r^2 d\Omega$. Hence show that

$$\frac{\mathbf{S}_{\text{out}} \cdot d\mathbf{\Sigma}}{S_{\text{in}}} = |f(\theta, \phi)|^2 d\Omega = d\sigma.$$

Solution: The probability current density is

$$\mathbf{S}(\mathbf{r}) = \frac{\hbar}{2\mu i} [u^* \nabla u - u \nabla u^*]$$

where

$$\nabla = \hat{r} \frac{\partial}{\partial r} + \hat{\theta} \frac{1}{r} \frac{\partial}{\partial \theta} + \hat{\phi} \frac{1}{r \sin \theta} \frac{\partial}{\partial \phi}$$

From the expression of u given in the problem, we note that

$$u_{out} \underset{r \to \infty}{\sim} Af(\theta, \phi)\frac{e^{ikr}}{r}, \quad u_{in} \underset{r \to \infty}{\sim} Ae^{ikz}$$

so that

$$\nabla u_{out} \sim A\hat{r}f(\theta, \phi)\frac{ik}{r}e^{ikr} + O\left(\frac{1}{r^2}\right)$$

$$\nabla u_{out}^* \sim A^*\hat{r}f^*(\theta, \phi)\frac{(-ik)}{r}e^{-ikr} + O\left(\frac{1}{r^2}\right)$$

Thus,

$$\mathbf{S}_{out} = \frac{\hbar|A|^2 2k}{2\mu r^2}|f(\theta, \phi)|^2\hat{r} + O\left(\frac{1}{r^3}\right)$$

Now

$$\frac{\hbar k}{\mu} = \frac{p}{\mu} = v$$

Hence

$$\mathbf{S}_{out} = \frac{v}{r^2}|A|^2|f(\theta, \phi)|^2\,\hat{r} + O\left(\frac{1}{r^3}\right)$$

Similarly,

$$\mathbf{S}_{in} = \frac{\hbar}{2\mu i}[u_{in}^*\nabla u_{in} - u_{in}\nabla u_{in}^*]$$

$$= \frac{\hbar}{2\mu i}\left[A^*e^{-ikz}\frac{\partial}{\partial z}Ae^{ikz} - Ae^{ikz}\frac{\partial}{\partial z}(A^*e^{-ikz})\right]$$

$$= \frac{\hbar}{2\mu i}|A|^2 2ki = v|A|^2$$

Now

$$\mathbf{S}_{out} \cdot d\mathbf{\Sigma} = \frac{v}{r^2}|A|^2|f(\theta, \phi)|^2\hat{r} \cdot d\mathbf{\Sigma} + O\left(\frac{1}{r^3}\right)$$

where

$$\hat{r} \cdot d\mathbf{\Sigma} = (d\mathbf{\Sigma})_r = r^2 d\Omega$$

Hence for large r

$$\frac{\mathbf{S}_{\text{out}} \cdot d\mathbf{\Sigma}}{S_{\text{in}}} = |f(\theta, \phi)|^2 d\Omega$$

$$= d\sigma$$

Q8.2 Consider the scattering of a beam of particles by a hard sphere of radius a:

$$V = \infty \quad r < a$$

$$= 0 \quad r > a.$$

Show that the phase shift δ_ℓ is given by

$$\tan \delta_\ell = \frac{j_\ell(ka)}{n_\ell(ka)};$$

where j_ℓ and n_ℓ have the usual meaning and k is the wave number. Find the cross section when $a \ll 1/k$. For $a \gg 1/k$, show that

$$\delta_\ell \approx -ka + \ell\pi/2.$$

Solution: The general solution of the free particle radial wave equation is

$$R_\ell(r) = A_\ell j_\ell(kr) + B_\ell n_\ell(kr), \quad r > a \qquad (8.1)$$

$$= 0 \qquad r < a \qquad (8.2)$$

The boundary condition at $r = a$ gives

$$A_\ell j_\ell(ka) + B_\ell n_\ell(ka) = 0 \qquad (8.3)$$

$$\frac{A_\ell}{B_\ell} = -\frac{n_\ell(ka)}{j_\ell(ka)} \qquad (8.4)$$

But from Eq. (8.1)

$$R_\ell(r) \underset{r \to \infty}{\sim} \frac{A_\ell}{kr} \sin\left(kr - \frac{1}{2}\ell\pi\right) - \frac{B_\ell}{kr} \cos\left(kr - \frac{1}{2}\ell\pi\right)$$

$$(8.5)$$

Put

$$A_\ell = C_\ell \cos \delta_\ell$$

$$B_\ell = -C_\ell \sin \delta_\ell \tag{8.6}$$

Then

$$R_\ell(r) \underset{r \to \infty}{\sim} \frac{C_\ell}{kr} \sin\left(kr - \frac{1}{2}\ell\pi + \delta_\ell\right)$$

From Eq. (8.6) the phase shift δ_ℓ is given by

$$\tan \delta_\ell = -\frac{B_\ell}{A_\ell}$$

$$= \frac{j_\ell(ka)}{n_\ell(ka)}, \quad \text{on using Eq. (8.4)} \tag{8.7}$$

Using Eqs. (8.85) of the text, we have for $ka \gg 1$:

$$\tan \delta_\ell = -\frac{\sin\left(ka - \frac{1}{2}\ell\pi\right)}{\cos\left(ka - \frac{1}{2}\ell\pi\right)}$$

$$= \tan\left(-ka + \frac{\ell\pi}{2}\right); \quad \delta_\ell = -ka + \frac{\ell\pi}{2}$$

whereas for $ka \ll 1$:

$$\tan \delta_\ell \sim \frac{-1}{(2\ell+1)!!} \frac{1}{(2\ell-1)!!}(ka)^{2\ell+1}$$

Here only s wave dominates ($\ell = 0$) for which $(2\ell+1)!! = 1$. Although $(2\ell - 1)!!$ is not defined for $\ell = 0$, $n_0(kr) = (-1)\frac{\cos kr}{kr} \underset{kr \to \infty}{\sim} -\frac{1}{kr}$ so that we can put $(-1)!! = 1$. Thus .

$$\tan \delta_0 \simeq -ka$$

$$\sigma = \frac{4\pi}{k^2} \sin^2 \delta_0 = 4\pi a^2, \tag{8.8}$$

independent of k.

Q8.3 Consider the scattering of a particle of mass m by a hard sphere of radius a

$$V = \infty \quad r < a$$

$$= 0 \quad r > a.$$

Treat the case for which the particle moves sufficiently slowly so that $\ell \geq 2$ phase shifts are negligible. Show that

$$\tan \delta_0 = -\tan(ka)$$

$$\tan \delta_1 = \frac{ka - \tan ka}{1 + ka \tan ka},$$

$k^2 = 2mE/\hbar^2$. If ka is small and the phase shifts δ_0 and δ_1 are small, show that

$$\delta_0 \approx -ka, \quad \delta_1 \approx -\frac{(ka)^3}{3}.$$

Show further that

$$\frac{d\sigma}{d\Omega} \sim a^2 \left(1 - \frac{(ka)^2}{3} + 2(ka)^2 \cos\theta + O(ka)^3 \right).$$

Solution: From Eq. (8.7)

$$\tan \delta_0 = \frac{j_0(ka)}{n_0(ka)} = \frac{\sin ka/ka}{-\cos ka/ka}$$

$$= -\tan ka \qquad (8.9)$$

$$\tan \delta_1 = \frac{j_1(ka)}{n_1(ka)} = \frac{-\frac{a}{k^2}\left[\frac{1}{r}\frac{d}{dr}\frac{\sin kr}{r}\right]_{r=a}}{-\frac{a}{k^2}\left[\frac{1}{r}\frac{d}{dr}\frac{\cos kr}{r}\right]_{r=a}}$$

$$= -\frac{-\frac{\sin(kr)}{r^2} + k\frac{\cos(kr)}{r}\big|_{r=a}}{-\frac{\cos(kr)}{r^2} - k\frac{\sin(kr)}{r}\big|_{r=a}}$$

$$\tan \delta_1 = \frac{ka - \tan ka}{1 + ka \tan ka} \qquad (8.10)$$

If $ka \ll 1$,

$$\tan ka = ka + \frac{1}{3}(ka)^3 \cdots \qquad (8.11)$$

Then from Eqs. (8.9), (8.10) and (8.11)

$$\delta_0 = -ka$$

and

$$\delta_1 \simeq \tan \delta_1 = \frac{ka - ka - \frac{1}{3}(ka)^3 \cdots}{1 + ka(ka + \frac{1}{3}(ka)^3 \cdots)} \qquad (8.12)$$

Hence

$$\delta_1 \approx -\frac{1}{3}(ka)^3$$

Now at low energies all the phase shifts for $\ell \geq 2$ are negligible.
Thus

$$
\begin{aligned}
\frac{d\sigma}{d\Omega} &= \frac{1}{k^2} \left| \sum_{\ell=0}^{\infty} (2\ell + 1) e^{i\delta_\ell} \sin \delta_\ell P_\ell(\cos \theta) \right|^2 \\
&= \frac{1}{k^2} |e^{i\delta_0} \sin \delta_0 + 3 e^{i\delta_1} \sin \delta_1 \cos \theta|^2 \\
&= \frac{1}{k^2} \{ \sin^2 \delta_0 + 9 \sin^2 \delta_1 \cos^2 \theta \\
&\qquad + 6 \sin \delta_0 \sin \delta_1 \cos (\delta_0 - \delta_1) \cos \theta \} \\
&\simeq \frac{1}{k^2} \left[\left(\delta_0 - \frac{1}{3!}\delta_0^3 \right)^2 + 6\delta_0\delta_1 \cos \theta \right] \\
&= \frac{\delta_0^2}{k^2} \left[1 - \frac{1}{3}\delta_0^2 + 6\frac{\delta_1}{\delta_0} \cos \theta \right] \\
&\sim a^2 \left[1 - \frac{1}{3}(ka)^2 + 2(ka)^2 \cos \theta \right] + O((ka)^4)
\end{aligned}
$$

Q8.4 Consider the scattering of a particle at low energies (i.e. small k, so that one can confine to s-wave only) by an attractive potential $V(r)$, the range of the potential being $r = a$. Suppose there exists among a discrete spectrum of negative energy levels a bound state (with angular momentum $\ell = 0$) at energy $E = -\epsilon(\epsilon > 0)$ where ϵ is very small. Show that the s-wave phase shift is given by

$$\tan \delta_0 = -\sqrt{E/\epsilon}$$

and

$$\sigma(E) = \frac{2\pi\hbar^2}{\mu(E + \epsilon)}.$$

Sketch $\sigma(E)$.

Solution: The scattering amplitude for s-wave

$$f = \frac{1}{k \cot \delta_0 - ik}$$

$$k \cot \delta_0 = -\frac{1}{a(E)}, \quad a \text{ is called the scattering length}$$

$$f = \frac{-1}{1/a + ik}$$

f has a pole at

$$ik = -\frac{1}{a} \quad k = i\kappa$$

$$\kappa = \frac{1}{a} \quad \text{or} \quad k \cot \delta_0 = -\kappa$$

This corresponds to a bound state if κ is positive and small.

$$\sigma = 4\pi|f|^2$$

$$= \frac{4\pi}{k^2} \frac{1}{\cot^2 \delta_0 + 1}$$

$$= 4\pi \frac{1}{1/a^2 + k^2}$$

$$= 4\pi \frac{1}{\kappa^2 + k^2}$$

$$= 4\pi \frac{\hbar^2}{2\mu(E + \epsilon)} \tag{8.13}$$

Q8.5 Use Born approximation to find the angular distribution (that is behaviour of $\sigma(\theta)$ with respect to θ) of the electrons in the elastic scattering of electrons by an atom represented by a shielded Coulomb potential

$$V(r) = -\frac{Ze^2}{r} \exp(-r/a).$$

Find also the total cross section.

Solution:

$$f_B(\theta) = -\frac{2\mu}{\hbar^2} \int_0^\infty \frac{\sin qr}{qr} V(r) r^2 dr$$

$$= \frac{2\mu}{\hbar^2} (Ze^2) \frac{1}{q} \mathrm{Im} \int_0^\infty \frac{1}{r} e^{iqr} e^{-r/a} dr$$

$$= \frac{2\mu Ze^2}{\hbar^2} \frac{1}{q} \mathrm{Im} \left\{ \frac{e^{(iq - \frac{1}{a})r}}{(iq - \frac{1}{a})} \bigg|_0^\infty \right\} \tag{8.14}$$

Now $e^{-r/a} \to 0$ as $r \to \infty$, so that

$$f_B(\theta) = \frac{2\mu Ze^2}{\hbar^2} \frac{1}{q} \mathrm{Im} \left(-\frac{1}{iq - \frac{1}{a}} \right)$$

$$= \frac{2\mu Ze^2}{\hbar^2} \frac{1}{q} \mathrm{Im} \left(-\frac{-iq - \frac{1}{a}}{q^2 + \frac{1}{a^2}} \right)$$

$$= \frac{2\mu Ze^2}{\hbar^2} \frac{1}{q^2 + \frac{1}{a^2}}$$

Now

$$\frac{d\sigma}{d\Omega} = |f_B(\theta)|^2$$

$$= \left(\frac{2\mu Ze^2}{\hbar^2}\right)^2 \frac{1}{\left[\frac{1}{a^2} + 4k^2 \sin^2 \frac{\theta}{2}\right]^2}$$

where $q = 2k \sin \frac{\theta}{2}$ and k is the wave number, $k^2 = \frac{2\mu E}{\hbar^2}$.

Q8.6 Consider the potential

$$V = V_0 \quad \text{for} \quad r < a$$

$$= 0 \quad \text{for} \quad r > a.$$

Find the scattering amplitude $f(\theta)$ in the Born approximation.

Solution: In the Born approximation for the central potential

$$f_B = -\frac{2\mu}{\hbar^2} \frac{1}{k} \int_0^\infty \sin(kr) r V(r) dr$$

For the potential

$$V(r) = V_0 \quad r < a$$

$$= 0 \quad r > a$$

$$f_B = -\frac{2\mu}{\hbar^2} \frac{V_0}{k} \int_0^a r \sin kr \, dr$$

where,

$$\int_0^a r \sin(kr) dr = -\frac{a}{k} \cos ka + \frac{1}{k^2} \sin ka$$

Hence

$$f_B = \frac{2\mu}{\hbar^2} \frac{V_0}{k^3} [ka \cos ka - \sin ka]$$

Q8.7 A solution of the equation

$$(\nabla^2 + k^2)G_k(\mathbf{r}) = \delta^3(\mathbf{r})$$

is given by

$$G_k(\mathbf{r}) = -(C_+ e^{ikr} + C_- e^{-ikr})\frac{1}{4\pi r},$$

where

$$C_+ + C_- = 1.$$

Verify that

$$\psi(\mathbf{r}) = e^{i\mathbf{k}\cdot\mathbf{r}} + \frac{2\mu}{\hbar^2}\int G_k(\mathbf{r} - \mathbf{r}')\psi(\mathbf{r}')V(\mathbf{r}')d^3r'$$

is solution of

$$\left(-\frac{\hbar^2}{2\mu}\nabla^2 + V(\mathbf{r})\right)\psi(\mathbf{r}) = E_k\psi(\mathbf{r}),$$

where

$$E_k = \frac{\hbar k^2}{2\mu^2}.$$

Show that in the limit of zero energy the asymptotic form
of ψ is

$$\psi \underset{r\to\infty}{\sim} 1 - \frac{a}{r},$$

where

$$a = \frac{\mu}{2\pi\hbar^2}\int \psi(\mathbf{r})V(\mathbf{r})d^3r.$$

Hence show that

$$\lim_{k \to 0} \delta_0/k = -a.$$

Solution:

$$\left[-\frac{\hbar^2}{2\mu} \nabla^2 + V(\mathbf{r}) \right] \psi(\mathbf{r}) = E_k \psi(\mathbf{r})$$

or

$$[\nabla^2 + k^2]\psi(\mathbf{r}) = \frac{2\mu}{\hbar^2} V(\mathbf{r})\psi(\mathbf{r})$$

Now

$$(\nabla^2 + k^2)\psi(\mathbf{r})$$

$$= (\nabla^2 + k^2)e^{i\mathbf{k}\cdot\mathbf{r}}$$

$$+ \frac{2\mu}{\hbar^2} \int \psi(\mathbf{r'})V(\mathbf{r'})(\nabla^2 + k^2)G_k(\mathbf{r} - \mathbf{r'})d^3r'$$

$$= 0 + \frac{2\mu}{\hbar^2} \int \psi(\mathbf{r'})V(\mathbf{r'})\delta^3(\mathbf{r} - \mathbf{r'})d^3r'$$

$$= \frac{2\mu}{\hbar^2} V(\mathbf{r})\psi(\mathbf{r}), \qquad (8.15)$$

giving the required verification. Now substituting the expression for $G_k(r)$ given in the problem

$$\psi(\mathbf{r}) = e^{i\mathbf{k}\cdot\mathbf{r}} - \frac{2\mu}{\hbar^2} \int \frac{1}{4\pi|\mathbf{r} - \mathbf{r'}|}(C_+ e^{ik|\mathbf{r}-\mathbf{r'}|} + C_- e^{-ik|\mathbf{r}-\mathbf{r'}|})$$

$$\times V(\mathbf{r'})\psi(\mathbf{r'})d^3r'$$

At low energies $k \to 0$

$$\psi(\mathbf{r}) \approx 1 - \frac{2\mu}{\hbar^2} \int \frac{1}{4\pi|\mathbf{r} - \mathbf{r'}|}[C_+ + C_-]V(\mathbf{r'})\psi(\mathbf{r'})d^3r'$$

For large r

$$|\mathbf{r} - \mathbf{r}'| = r - \hat{\mathbf{r}} \cdot \mathbf{r}' \qquad (8.16)$$

Thus at low energies

$$\psi(\mathbf{r}) \xrightarrow[r \to \infty]{} 1 - \frac{2\mu}{\hbar^2} \frac{1}{4\pi r} \int V(\mathbf{r}')\psi(\mathbf{r}')d^3 r'$$

$$= \left(1 - \frac{a}{r}\right) \qquad (8.17)$$

where

$$a = \frac{\mu}{2\pi\hbar^2} \int V(\mathbf{r}')\psi(\mathbf{r}')d^3 r'$$

Now

$$\psi(\mathbf{r}) \xrightarrow[r \to \infty]{} e^{i\mathbf{k}\cdot\mathbf{r}} + f(\theta, \phi)\frac{e^{ikr}}{r}$$

$$\rightarrow \quad 1 + \frac{f(\theta, \phi)}{r}(1 + ikr \cdots) \qquad (8.18)$$

Thus (for $k \to 0$, only s-wave dominates)

$$a = -f(\theta) = -\frac{1}{k}e^{i\delta_0}\sin\delta_0$$

$$\approx -\frac{1}{k}\delta_0 \qquad (8.19)$$

Q8.8 The scattering amplitude in terms of phase shifts is given by

$$f(\theta) = \frac{1}{2ik}\sum_{l=0}^{\infty}(2l + 1)(e^{2i\delta_\ell} - 1)P_\ell(\cos\theta). \qquad (8.20)$$

The Born formula for the scattering amplitude $f(\theta)$ is

$$f_B(\theta) = -\frac{2\mu}{\hbar^2}\int_0^\infty \frac{\sin qr}{qr}V(r)r^2 dr.$$

For small phase shifts, $e^{2i\delta_\ell} - 1 \approx 2i\delta_\ell$. Using the formula

$$\frac{\sin qr}{qr} = \sum_\ell (2\ell + 1)j_\ell^2(kr)P_\ell(\cos\theta),$$

show that for small δ_ℓ:

$$\delta_\ell^B \approx \frac{2\mu}{\hbar^2}k \int_0^\infty V(r)j_\ell^2(kr)r^2 dr.$$

Since for $\ell > \ell_{\max}$, the phase shifts are small, we can use the Born amplitude as a device for summing up the partial wave series for $\ell > \ell_{\max}$. Using the above statement show that

$$f(\theta) = \frac{1}{2ik}\sum_{\ell=0}^{\ell_{\max}}(2\ell + 1)((e^{2i\delta_\ell} - 1)$$

$$- (e^{2i\delta_\ell^B} - 1))P_\ell(\cos\theta) + f_B(\theta).$$

Solution:

$$f_B(\theta) = \sum_\ell \left[-\frac{2\mu}{\hbar^2}\int_0^\infty V(r)r^2 j_\ell^2(kr)dr\right](2\ell + 1)P_\ell(\cos\theta)$$

But from Eq. (8.20), for small phase shifts

$$f_B(\theta) \approx \frac{1}{2ik}\sum_\ell(2\ell + 1)(2i\delta_\ell)P_\ell(\cos\theta)$$

Thus for small phase shift, we have

$$\delta_\ell^B = -\frac{2\mu}{\hbar^2}k\int_0^\infty V(r)j_\ell^2(k\ell)r^2 dr$$

We can write

$$f(\theta) = \frac{1}{2ik}\left[\sum_{\ell=0}^{\ell_{max}}(2\ell + 1)(e^{2i\delta_\ell} - 1)P_\ell(\cos\theta)\right.$$

$$\left. + \sum_{\ell_{max}}^\infty(2\ell + 1)(e^{2i\delta_\ell} - 1)P_\ell(\cos\theta)\right]$$

$$\approx \frac{1}{2ik} \left[\sum_{\ell=0}^{\ell_{max}} (2\ell + 1)(e^{2i\delta_\ell} - 1)P_\ell(\cos\theta) \right.$$

$$\left. + \sum_{\ell_{max}}^{\infty} (2\ell + 1)(2i\delta_\ell^B)P_\ell(\cos\theta) \right]$$

Now

$$\sum_{\ell_{max}}^{\infty} = \sum_{\ell=0}^{\infty} - \sum_{\ell}^{\ell_{max}}$$

Thus we can write

$$f(\theta) = \frac{1}{2ik} \sum_{\ell=0}^{\ell_{max}} (2\ell + 1)[(e^{2i\delta_\ell} - 1)$$

$$- (e^{2i\delta_\ell^B} - 1)]P_\ell(\cos\theta) + f_B(\theta)$$

Q8.9 Consider the scattering of a particle of mass m by the potential

$$V = V_0 \quad r < a$$

$$= 0 \quad r > a.$$

Estimate the differential scattering cross section for $\hbar^2 k^2 \gg 2mV_0$ and discuss the validity of the approximation used.

In the limit $ka \ll 1$, show that the cross section takes on the form

$$\frac{d\sigma}{d\Omega} = A + B\cos\theta$$

with $B \ll A$.

Solution:

$$f_B(\theta) = -\frac{2\mu}{\hbar^2} \int_0^\infty \frac{\sin qr}{qr} V(r)r^2 dr$$

$$= -\frac{2\mu V_0}{\hbar^2} \int_0^a \frac{\sin qr}{q} r dr$$

.

Integrating by parts,

$$f_B(\theta) = -\frac{2\mu V_0}{\hbar^2} \left[\frac{\sin qa - qa \cos qa}{q^3} \right]$$

Thus

$$\frac{d\sigma}{d\Omega} = |f_B(\theta)|^2$$

$$= \left(\frac{2\mu V_0}{\hbar^2} \right)^3 \frac{1}{q^6} (\sin qa - qa \cos qa)^2 \qquad (8.21)$$

Put $x = qa$, $q = 2k \sin \frac{\theta}{2}$, $ka \ll 1$ because $x \ll 1$

$$\frac{d\sigma}{d\Omega} = \left(\frac{2\mu V_0}{\hbar^2} \right)^3 \frac{a^6}{x^6} [\sin x - x \cos x]^2$$

Using the expressions

$$\sin x = x - \frac{1}{3!} x^3 + \frac{1}{5!} x^5 + \cdots$$

$$x \cos x = x \left[1 - \frac{1}{2!} x^2 + \frac{1}{4!} x^4 \cdots \right] \qquad (8.22)$$

$$\frac{d\sigma}{d\Omega} = \left(\frac{2\mu V_0 a^2}{\hbar^2} \right)^3 \frac{1}{9} \left[1 - \frac{1}{5} x^2 \right]$$

$$= \left(\frac{2\mu V_0 a^2}{\hbar^2} \right)^3 \frac{1}{9} \left[1 - \frac{2a^2}{5} k^2 (1 - \cos \theta) \right]$$

$$= \left(\frac{2\mu V_0 a^2}{\hbar^2} \right)^3 \frac{1}{9} \left[1 - \frac{2}{5} k^2 a^2 + \frac{2k^2 a^2}{5} \cos \theta \right]$$

$$= A + B \cos \theta \qquad (8.23)$$

where

$$\frac{B}{A} \approx \frac{2k^2 a^2}{5} \ll 1$$

Q8.10 Consider an electron confined inside a sphere of radius a. What is the pressure exerted on the surface of the sphere, if the electron is in (i) the lowest s state (ii) the lowest p-state?

$$\text{Hint:} \qquad V = 0 \qquad r < a$$
$$= \infty \quad r > a.$$

and $P = -\frac{\partial E}{\partial V}$, [$V$ here is volume].

Solution: For s-wave, the radial Schrödinger equation in the region $r < a$ is

$$\frac{d^2\chi_0}{dr^2} + k^2\chi_0 = 0$$

where $k = \sqrt{2mE}/\hbar$. This has the solution

$$\chi_0(r) = A_1 e^{ikr} + A_2 e^{-ikr}$$
$$= A \sin kr + B \cos kr$$

The boundary condition $\chi_0(0) = 0$, gives $B = 0$

$$\chi_0(r) = A \sin kr$$
$$R_0(r) = A \sin\left(\frac{kr}{r}\right) \qquad (8.24)$$

The boundary condition at $r = a$

$$R(a) = 0$$

gives

$$\sin ka = 0; \Rightarrow ka = n\pi \qquad (8.25)$$

or

$$k^2 a^2 = n^2 \pi^2$$

Hence

$$E_n = \frac{n^2 \pi^2 \hbar^2}{2ma^2}$$

For the lowest s-state $(n = 1)$

$$E = \frac{\pi^2 \hbar^2}{2ma^2}$$

The volume of the sphere of radius a is

$$V = \frac{4\pi}{3} a^3$$

Hence

$$P = -\frac{\partial E}{\partial V}$$

$$= -\frac{\partial E}{\partial a} \frac{\partial a}{\partial V}$$

$$= \frac{\pi \hbar^2}{4ma^5}$$

For p-wave, we have

$$\frac{d^2 \chi}{dr^2} + \left[k^2 - \frac{2}{r^2} \right] \chi = 0 \quad r < a$$

Now the general solution of the Schrödinger equation for $V(r) = 0$

$$\frac{d^2 R_\ell}{dr^2} + \frac{2}{r} \frac{dR_\ell}{dr} + \left[k^2 - \frac{\ell(\ell+1)}{r^2} \right] R_\ell(r) = 0 \quad r < a$$

which is regular at $r = 0$ and is given by

$$R_\ell(r) = A_\ell j_\ell(kr)$$

For p-wave, the solution is

$$R_1(r) = A_1 j_1(kr)$$

$$= A_1 \left[\frac{\sin kr}{k^2 r^2} - \frac{\cos kr}{kr} \right] \tag{8.26}$$

The boundary condition $R_1(a) = 0$, gives

$$\frac{\sin ka}{k^2 a^2} - \frac{\cos ka}{ka} = 0$$

Hence the energy eigenvalues of the p-state are given by

$$ka \cot ka = 1$$

which can be solved numerically. Once the energy is determined numerically for the lowest p-state, the pressure can be easily calculated.

Q8.11 Show that for *s*-wave bound states, the wave function at the origin is given by

$$|\psi_s(0)|^2 = \frac{2\mu}{4\pi} \left\langle \frac{dV}{dr} \right\rangle,$$

where V is the potential between two particles and μ is their reduced mass.

Solution: For *s*-wave, we have $\psi_s(\mathbf{r}) = R(r)Y_{00} = (1/\sqrt{4\pi})R(r) = (1/\sqrt{4\pi})\frac{\chi(r)}{r}$, where $\chi(r)$ satisfies the radial Schrödinger equation,

$$\frac{d^2\chi}{dr^2} + \left[\frac{2\mu}{\hbar^2}(E - V(r)) \right] \chi = 0$$

Writing $\chi'' = \frac{d^2\chi}{dr^2}$, we have

$$\frac{d}{dr}(\chi''/\chi) = \frac{2\mu}{\hbar^2}\frac{dV}{dr}$$

For a bound state, the wave function $\chi(r)$ is real. From the above equation, we have

$$\int_0^\infty \chi \left[\frac{d}{dr}(\chi''/\chi) \right] \chi dr = \frac{2\mu}{\hbar^2} \int_0^\infty \chi \frac{dV}{dr} \chi dr.$$

Integrating the left-hand side by parts, we have

$$\int_0^\infty \chi[d/dr(\chi''/\chi)]\chi dr = \chi\chi''|_0^\infty - (\chi')^2|_0^\infty.$$

Using the boundary conditions for a bound state wave function $R(r)$;

$$R(r) \xrightarrow[r \to 0]{} R(0), \qquad \text{finite}$$

$$\chi(r) \xrightarrow[r \to 0]{} 0, \qquad \chi(r) \xrightarrow[r \to \infty]{} 0$$

and $\chi'(r) = R(r) + rR'(r)$ we have

$$\int_0^\infty \chi[d/dr(\chi''/\chi)]\chi dr = |R(0)|^2.$$

Hence

$$|R(0)|^2 = (2\mu/\hbar^2) \int_0^\infty R(r)(dV/dr)R(r)r^2 dr$$

$$|\psi_s(0)|^2 = \left(\frac{2\mu}{4\pi\hbar^2} \right) \int \psi_s(\mathbf{r}) \frac{dV}{dr} \psi_s(\mathbf{r}) d\mathbf{r}$$

or

$$|\psi_s(0)|^2 = (2\mu/4\pi\hbar^2)\langle dV/dr \rangle.$$

This result is useful for the interpretation of the J/ψ at 3100 MeV. This resonance is regarded as a s-wave bound state of charmed quark and antiquark ($c\bar{c}$ called charmonium). It is possible to determine $|\psi_s(0)|^2$ from the decay rate $J/\psi \rightarrow e^- e^+$. Then using the above relation, we can get information about the potential $V(r)$, which keeps $c\bar{c}$ bound together. Since the mass of charmed quark is high (1500 MeV–2000 MeV), it is a good approximation to treat the $c\bar{c}$ system non-relativistically with $\mu = \frac{1}{2}m_c$.

Q8.12 Using the WKB approximation, show that for the s-wave bound states for the logarithmic potential $V(r) = C \ln r/r_0$, the energy eigenvalues are given by

$$E_n = C \ln \left[\left(2n - \frac{1}{2} \right) \sqrt{\pi} \right] + C \ln[(\hbar/\sqrt{m_c C})1/r_0],$$

where m_c is the mass of charmed quark. Show that the mean square radius of the charmonium for the logarithmic potential is given by

$$\langle \psi_{ns} | r^2 | \psi_{ns} \rangle = (1/C m_c)(4\pi/\sqrt{3})(n - 1/4)^2.$$

Hint: The quantization condition is

$$\int_0^{r_1} [m_c(E - C \ln(r/r_0))]^{1/2} dr = (n - 1/4)\pi\hbar,$$

where $r_1 = r_0 e^{E/C}$. In order to perform the integration, put $(E/C - y^2) = \ln(r/r_0)$.

Solution: In the WKB approximation, the bound state wave function $\chi(r)$ for s-wave is given by

$$\chi(r) = \frac{N}{\sqrt{k(r)}} \sin \left[\int_r^{r_1} k(r') dr' + \frac{\pi}{4} \right] \qquad (8.27)$$

where r_1 is the first turning point $(k(r_i) = 0)$. The boundary condition $\chi(0) = 0$ gives the quantization condition

$$\int_0^{r_1} k(r) dr + \frac{\pi}{4} = n\pi$$

or

$$\int_0^{r_1} \left[\frac{2\mu}{\hbar^2} (E - V(r)) \right]^{1/2} dr = \left(n - \frac{1}{4} \right) \pi \qquad (8.28)$$

where $n = 1, 2, 3, \ldots$. For the logarithmic potential

$$V(r) = C \ln \left(\frac{r}{r_0} \right)$$

Thus the first turning point r_1 is given by

$$E = \ln \left(\frac{r_1}{r_0} \right) \Rightarrow \frac{r_1}{r_0} = e^{E/C}$$

Now $[\mu = \frac{1}{2} m_c]$

$$\frac{\sqrt{2\mu}}{\hbar} \int_0^{r_1} [E - C \ln(r/r_0)]^{1/2} dr$$

$$= \sqrt{\frac{m_c}{\hbar^2}} (2r_0) e^{E/C} \sqrt{C} \int_0^{\infty} y^2 e^{-y^2} dy$$

$$= \frac{\sqrt{\pi}}{2} \sqrt{\frac{2m_c C}{\hbar^2}} r_0 e^{E/C} \qquad (8.29)$$

where we have put

$$r = r_0 e^{E/C} e^{-y^2}$$

so that

$$\ln r/r_0 = E/C - y^2$$
$$dr = r_0 e^{E/C} (-2y) e^{-y^2}$$
$$r = 0, \qquad y \to \infty$$
$$r = r_1, \qquad y = 0$$

Hence from Eqs. (8.28) and (8.29)

$$e^{E/C} = \frac{2\sqrt{\pi\hbar^2}}{\sqrt{m_c C}} \frac{1}{r_0} \left(n - \frac{1}{4} \right) \tag{8.30}$$

$$E = C \ln \left[\left(2n - \frac{1}{2} \right) \sqrt{\pi} \right] - C \ln(r_0 \sqrt{m_c C / \hbar^2})$$

Now

$$\langle r^2 \rangle_{ns} = \int_0^{r_1} |\psi_{ns}|^2 r^4 dr d\Omega$$

$$= \int_0^{r_1} |\chi_{ns}|^2 r^2 dr$$

$$= \int_0^{r_1} \frac{|N|^2 r^2}{k(r)} \left\{ \sin \left[\int_r^{r_1} k(r') dr' + \frac{\pi}{4} \right] \right\}^2 dr$$

In order to evaluate the integral, we replace the $\sin^2$ term by its average value of $1/2$.

$$\langle r^2 \rangle_{ns} = |N|^2 \frac{1}{2} \int_0^{r_1} \frac{r^2}{k(r)} dr$$

$$= \frac{|N|^2}{2\sqrt{m_c C}/\hbar} \int_0^{r_1} \frac{r^2 dr}{[E - C \ln r/r_0]^{1/2}}$$

$$= \frac{|N|^2 2}{2\sqrt{m_c C}/\hbar} r_0^3 e^{3E/C} \int_0^\infty e^{-3y^2} dy$$

$$= \frac{|N|^2}{\sqrt{m_c C}/\hbar} r_0^3 e^{3E/C} \frac{\sqrt{\pi}}{2\sqrt{3}} \tag{8.31}$$

The normalization condition:

$$N^2 \int_0^{r_1} \chi_{ns}^2 dr = 1 \tag{8.32}$$

Now

$$\int_0^{r_1} \chi_{ns}^2 dr = \frac{1}{2} \int_0^{r_1} \frac{dr}{k(r)} = \int_0^{r_1} \frac{1}{2} \frac{dr}{\sqrt{m_c(E - V)}} \tag{8.33}$$

From Eq. (8.28)

$$\frac{1}{2}\sqrt{\frac{m_c}{\hbar^2}}\int_0^{r_1}\frac{dr}{(E-V(r))^{1/2}}\frac{\partial E}{\partial n}=\pi$$

or

$$\int_0^{r_1}\frac{dr}{[m_c(E-V(r))]^{1/2}}=\frac{2\pi\hbar}{m_c}\frac{1}{(\partial E/\partial n)}$$

Hence from Eq. (8.27), using Eqs. (8.33) and (8.32), we have

$$N^2=\frac{m_c}{\pi\hbar^2}\left(\frac{\partial E}{\partial n}\right) \tag{8.34}$$

Now from Eq. (8.30)

$$\frac{\partial E}{\partial n}=2\sqrt{\frac{\pi C}{m_c}}\frac{\hbar}{r_0}e^{-E/C}$$

Thus

$$N^2=2\sqrt{\frac{m_cC}{\pi\hbar^2}}\frac{1}{r_0}e^{-E/C}$$

On using Eq. (8.31).

$$\langle r^2\rangle_{ns}=\frac{r_0^2}{\sqrt{3}}e^{2E/C}$$

Hence using Eq. (8.31)

$$\langle r^2\rangle_{ns}=\frac{4\pi}{\sqrt{3}}\frac{\hbar^2}{m_cC}\left(n-\frac{1}{4}\right)^2$$

Chapter 9

Operators

Q9.1 $\hat{A}$ and $\hat{B}$ are two arbitrary hermitian operators. Which of the following operators

(i) $\hat{A}\hat{B}$, (ii) $\hat{A}^2$, (iii) $\hat{A}\hat{B} - \hat{B}\hat{A}$, (iv) $\hat{A}\hat{B} + \hat{B}\hat{A}$, (v) $\hat{A}\hat{B}\hat{A}$

(a) are hermitian,

(b) have real non-negative expectation values,

(c) have pure imaginary expectation values,

(d) are purely numerical operators.

Solution: (a)

$$\langle\phi|(\hat{A}\hat{B})|\psi\rangle^* = \langle\psi|(\hat{A}\hat{B})^\dagger|\phi\rangle \tag{9.1}$$

But

$$\langle\phi|\hat{A}\hat{B}|\psi\rangle^* = \sum_n \langle\phi|\hat{A}|n\rangle^* \langle n|\hat{B}|\psi\rangle^*$$

$$= \sum_n \langle\psi|\hat{B}^\dagger|n\rangle\langle n|\hat{A}^\dagger|\phi\rangle$$

$$= \langle\psi|\hat{B}^\dagger\hat{A}^\dagger|\phi\rangle = \langle\psi|\hat{B}\hat{A}|\phi\rangle \tag{9.2}$$

Hence

$$\langle\psi|(\hat{A}\hat{B})^\dagger|\phi\rangle = \langle\psi|\hat{B}\hat{A}|\phi\rangle$$

Since $|\psi\rangle$ and $|\phi\rangle$ are arbitrary states,

$$(\hat{A}\hat{B})^\dagger = \hat{B}^\dagger\hat{A}^\dagger = \hat{B}\hat{A} \tag{9.3}$$

(i) $(\hat{A}\hat{B})$ is not hermitian.

(ii) $\hat{A}^2 = \hat{A}^\dagger\hat{A}^\dagger = \hat{A}^2$, hermitian

(iii)

$$(\hat{A}\hat{B} - \hat{B}\hat{A})^\dagger = (\hat{B}\hat{A} - \hat{A}\hat{B})$$
$$= -(\hat{A}\hat{B} - \hat{B}\hat{A}), \qquad \text{anti-hermitian}$$

(iv)

$$(\hat{A}\hat{B} + \hat{B}\hat{A})^\dagger = (\hat{A}\hat{B} + \hat{B}\hat{A}), \quad \text{hermitian}$$

(v)

$$(\hat{A}\hat{B}\hat{A})^\dagger = (\hat{F}\hat{A})^\dagger \quad \text{where } \hat{F} = \hat{A}\hat{B}$$
$$= \hat{A}^\dagger\hat{F}^\dagger = \hat{A}^\dagger\hat{B}^\dagger\hat{A}^\dagger = \hat{A}\hat{B}\hat{A}, \quad \text{hermitian}$$

(b) (i)

$$\langle\psi|\hat{A}\hat{B}|\psi\rangle^* = \langle\psi|(\hat{A}\hat{B})^\dagger|\psi\rangle = \langle\psi|\hat{B}\hat{A}|\psi\rangle$$

i.e. average value is complex.

(ii)

$$\langle\psi|\hat{A}^2|\psi\rangle^* = \langle\psi|\hat{A}^2|\psi\rangle$$

$\hat{A}^2$ has real expectation value.

Now

$$\langle\psi|\hat{A}^2|\psi\rangle = \sum_n \langle\psi|\hat{A}|n\rangle\langle n|\hat{A}|\psi\rangle$$

$$= \sum_n |\langle n|\hat{A}|\psi\rangle|^2 \tag{9.4}$$

$$\geq 0 \tag{9.5}$$

$\hat{A}^2$ has real (non-negative) average value.

(iii)

$$\langle\psi|(\hat{A}\hat{B} - \hat{B}\hat{A})|\psi\rangle^* = -\langle\psi|(\hat{A}\hat{B} - \hat{B}\hat{A})|\psi\rangle$$

i.e. the average value is purely imaginary.

(iv)

$$\langle \psi|(\hat{A}\hat{B} + \hat{B}\hat{A})|\psi\rangle^* = \langle \psi|(\hat{A}\hat{B} + \hat{B}\hat{A})|\psi\rangle$$

average value is real.

(v)

$$\langle \psi(\hat{A}\hat{B}\hat{A})\psi\rangle^* = \langle \psi(\hat{A}\hat{B}\hat{A})^\dagger\psi\rangle$$
$$= \langle \psi\hat{A}\hat{B}\hat{A}\psi\rangle \qquad (9.6)$$

average value is real.

Summary: (a) $\hat{A}^2, (\hat{A}\hat{B} + \hat{B}\hat{A})$, $\hat{A}\hat{B}\hat{A}$ are hermitian.

(b) Only $\hat{A}^2$ has real non-negative average value.

(c) Only $(\hat{A}\hat{B} - \hat{B}\hat{A})$ has purely imaginary average value.

Q9.2 If an operator $\hat{A}$ has the property that

$$\hat{A}^4 = 1.$$

What are its eigenvalues?

Solution: $\widehat{A}^2$ have eigenvalues as solution of

$$(\widehat{A}^2 - 1)(\widehat{A}^2 + 1) = 0$$

i.e. $\widehat{A}^2$ has eigenvalue $+1$ or -1. For $\widehat{A}^2 = 1$, $\widehat{A}$ has eigenvalues ± 1. For $\widehat{A}^2 = -1$, $\widehat{A}$ has eigenvalues $\pm i$.

Q9.3 The Hamiltonian

$$H = \frac{\hat{p}^2}{2m} + V(\hat{x})$$

has a set of eigenstates $|n\rangle$ with energy eigenvalues E_n. The lowest eigenstate $|0\rangle$ has the energy E_0. Show that

$$\sum_n (E_n - E_0)|\langle n|x|0\rangle|^2 = \frac{\hbar^2}{2m}.$$

Hint:

$$[\hat{x}, \hat{p}] = i\hbar$$

$$[\hat{x}, [H, \hat{x}]] = \frac{\hbar^2}{m}.$$

Verify the above sum rule for a simple harmonic oscillator.

Solution: Now

$$[H, \hat{x}] = \left[\frac{\hat{p}^2}{2m}, \hat{x} \right] + [V(\hat{x}), \hat{x}]$$

$$= \frac{1}{2m}[\hat{p}^2, \hat{x}]$$

$$= \frac{1}{2m}\{\hat{p}[\hat{p}, \hat{x}] + [\hat{p}, \hat{x}]\hat{p}\}$$

$$= -\frac{i\hbar}{m}\hat{p}$$

Thus

$$[\hat{x}, [H, \hat{x}]] = -\frac{i\hbar}{m}[\hat{x}, \hat{p}]$$

$$= \frac{\hbar^2}{2m} \tag{9.7}$$

Taking the matrix elements between the states $\langle m|$ and $|0\rangle$ of Eq. (9.7) we have

$$\langle m|[\hat{x}, [H, \hat{x}]]|0\rangle = \frac{\hbar^2}{2m}\delta_{m0} \tag{9.8}$$

Now

$$\langle m|[\hat{x}, [H, \hat{x}]]|0\rangle = \sum_n \{\langle m|\hat{x}|n\rangle\langle n|[H, \hat{x}]|0\rangle$$

$$-\langle m|[H, \hat{x}]|n\rangle\langle n|\hat{x}|0\rangle\}$$

$$= \sum_n \{\langle m|\hat{x}|n\rangle(E_n - E_0)\langle n|\hat{x}|0\rangle$$

$$-(E_m - E_n)\langle m|\hat{x}|n\rangle\langle n|\hat{x}|0\rangle\} \tag{9.9}$$

Hence on using Eq. (9.9) for $m = 0$, we get from Eq. (9.8)

$$\sum_n [(E_n - E_0)\langle 0|\hat{x}|n\rangle\langle n|\hat{x}|0\rangle$$

$$+ (E_n - E_0)\langle 0|\hat{x}|n|n\rangle\langle n|\hat{x}|0\rangle] = \frac{\hbar^2}{m}$$

or

$$\sum_n [(E_n - E_0) \, |\langle n|\hat{x}|0\rangle|^2] = \frac{\hbar^2}{2m} \qquad (9.10)$$

For a simple harmonic oscillator

$$\hat{x} = \sqrt{2m\hbar\omega} \, \frac{1}{2i\omega m}(a^\dagger - a)$$

$$= \frac{i\sqrt{\hbar}}{\sqrt{2m\omega}}(a - a^\dagger)$$

$$\langle n|\hat{x}|0\rangle = \frac{i\sqrt{\hbar}}{\sqrt{2m\omega}}\langle n|(a - a^\dagger)|0\rangle$$

$$= -\frac{i\sqrt{\hbar}}{\sqrt{2m\omega}}\langle n|1\rangle$$

$$= -\frac{i\sqrt{\hbar}}{\sqrt{2m\omega}}\delta_{n1} \qquad (9.11)$$

Hence L.H.S of Eq. (9.10)

$$= \sum_n (E_n - E_0)\frac{\hbar}{2m\omega}\delta_{n1}\delta_{n1}$$

$$= (E_1 - E_0)\frac{\hbar}{2m\omega} = \frac{\hbar\omega\hbar}{2m\omega} = \frac{\hbar^2}{2m} = \text{R.H.S}$$

Q9.4 In the momentum representation, show that the position operator $\hat{x}$ is represented by

$$\langle p|\hat{x}|p'\rangle = i\hbar\frac{\partial}{\partial p}\langle p|p'\rangle$$

$$\langle p|\hat{x}|\psi\rangle = i\hbar\frac{\partial}{\partial p}\langle p|\psi\rangle$$

that is

$$\hat{x} = i\hbar\frac{\partial}{\partial p}.$$

Solution: To find the position operator $\hat{x}$ in momentum representation we use $\hat{x}\hat{p} - \hat{p}\hat{x} = i\hbar$

$$\langle p|\hat{x}\hat{p} - \hat{p}\hat{x}|p'\rangle = i\hbar\langle p|p'\rangle$$

$$= i\hbar\delta(p - p') \qquad (9.12)$$

$$\hat{p}|p'\rangle = p'|p'\rangle$$

$$\langle p|\hat{p} = p\langle p|$$

On using the above equations, we have from Eq. (9.12)

$$(p' - p)\langle p|\hat{x}|p'\rangle = i\hbar\delta(p - p') \qquad (9.13)$$

Now

$$(p - p')\frac{\partial}{\partial p}\delta(p - p') = -\delta(p - p')$$

[cf. Eq. (9.41d) of the text with x replaced by p]. Thus from Eq. (9.13), we get

$$(p' - p)\langle p|\hat{x}|p'\rangle = (p' - p)(-i\hbar)\frac{\partial}{\partial p}\delta(p - p')$$

Thus [interchanging $p \to p'$]

$$\langle p'|\hat{x}|p\rangle = -i\hbar\frac{\partial}{\partial p'}\delta(p - p')$$

$$= \delta(p - p')i\hbar\frac{\partial}{\partial p} \qquad (9.14)$$

Now

$$\langle p|\hat{x}|\psi\rangle = \int dp' \langle p|\hat{x}|p'\rangle\langle p'|\psi\rangle$$

$$= \int dp' \delta(p' - p)i\hbar\frac{\partial}{\partial p}\langle p'|\psi\rangle$$

$$= i\hbar\frac{\partial}{\partial p}\langle p|\psi\rangle,$$

Thus

$$\hat{x} \to i\hbar\frac{\partial}{\partial p}$$

Q9.5 If $|n\rangle$ denotes a normalised eigenstate of a simple harmonic oscillator, belonging to the eigenvalue $E_n = (n + \frac{1}{2})\hbar\omega$, show that

$$\langle n|\hat{x}|m\rangle = \begin{cases} i\sqrt{\hbar/m\omega}\left(\dfrac{n+1}{2}\right)^{1/2}, & m = n+1 \\[2ex] -i\sqrt{\hbar/m\omega}\left(\dfrac{n}{2}\right)^{1/2}, & m = n-1 \\[2ex] 0 & \text{otherwise.} \end{cases}$$

(9.15)

Solution: Now

$$\hat{x} = i\sqrt{\frac{\hbar}{2m\omega}}(a - a^\dagger)$$

$$\hat{p} = \sqrt{\frac{m\hbar\omega}{2}}(a + a^\dagger)$$

Therefore

$$\langle n|\hat{x}|m\rangle = i\sqrt{\frac{\hbar}{2m\omega}}\left[\sqrt{m}\delta_{n,m-1} - \sqrt{m+1}\delta_{n,m+1}\right]$$

(9.16)

$$\langle n|\hat{p}|m\rangle = \sqrt{\frac{m\hbar\omega}{2}}\left[\sqrt{m}\delta_{n,m-1} + \sqrt{m+1}\delta_{n,m+1}\right]$$

(9.17)

where we have used

$$\hat{a}^\dagger|m\rangle = \sqrt{m+1}|m+1\rangle$$
$$\hat{a}|m\rangle = \sqrt{m}|m-1\rangle$$

$$\therefore \langle n|\hat{x}|m\rangle = \begin{cases} i\sqrt{\hbar/m\omega}\left(\dfrac{n+1}{2}\right)^{1/2}, & m = n+1 \\[2ex] -i\sqrt{\hbar/m\omega}\left(\dfrac{n}{2}\right)^{1/2}, & m = n-1 \\[2ex] 0 & \text{otherwise} \end{cases}$$

Since $\langle n|\hat{x}|n\rangle = 0$ and $\langle n|\hat{p}|n\rangle = 0$, there is no contradiction in having i in the matrix elements $\langle n|\hat{x}|m\rangle$.

Similarly

$$\langle n|\hat{p}|m\rangle = \begin{cases} \sqrt{\hbar/m\omega}\left(\dfrac{n+1}{2}\right)^{1/2}, & m = n+1 \\[3mm] \sqrt{\hbar/m\omega}\left(\dfrac{n}{2}\right)^{1/2}, & m = n-1 \\[3mm] 0 & \text{otherwise.} \end{cases}$$

Q9.6 If an operator $\hat{A}$ has the following properties

$$\hat{A}^2 = 0$$
$$\hat{A}\hat{A}^\dagger + \hat{A}^\dagger\hat{A} = 1,$$

show that

(i) $N = \hat{A}^\dagger\hat{A}$ is hermitian.
(ii) $N^2 = N$. Hence show that N has eigenvalues 0 and 1.
(iii) $[N, \hat{A}^\dagger] = \hat{A}^\dagger$.
(iv) Let $|0\rangle$ and $|1\rangle$ be eigenvectors of N belonging to eigenvalues 0 and 1. Show that

$$\hat{A}^\dagger|0\rangle = 1$$
$$\hat{A}^\dagger|1\rangle = 0.$$

Solution:

(i)

$$N = \hat{A}^\dagger\hat{A}$$
$$N^\dagger = \hat{A}^\dagger\hat{A} = N$$

Hence N is hermitian.

(ii)

$$N^2 = \hat{A}^\dagger\hat{A}\hat{A}^\dagger\hat{A} = \hat{A}^\dagger\hat{A}(1 - \hat{A}\hat{A}^\dagger) = \hat{A}^\dagger\hat{A} = N$$

Thus $N(N-1) = 0$

Hence eigenvalues of N are 0 and 1.

(iii)

$$[N, \hat{A}^\dagger] = [\hat{A}^\dagger \hat{A}, \hat{A}^\dagger] = \hat{A}^\dagger \hat{A} \hat{A}^\dagger - \hat{A}^\dagger \hat{A}^\dagger A = \hat{A}^\dagger \hat{A} \hat{A}^\dagger$$
$$= \hat{A}^\dagger (1 - \hat{A}^\dagger \hat{A}) = \hat{A}^\dagger$$

(iv)

$$N|0\rangle = 0, \quad N|1\rangle = |1\rangle$$

Now on using (iii)

$$N\hat{A}^\dagger|0\rangle = (\hat{A}^\dagger + \hat{A}^\dagger N)|0\rangle = \hat{A}^\dagger|0\rangle$$

Thus $\hat{A}^\dagger|0\rangle$ is either zero or an eigenstate of N with eigenvalue C. Hence

$$\hat{A}^\dagger|0\rangle = C|1\rangle$$

Now

$$|C|^2 = \langle 0|\hat{A}\hat{A}^\dagger|0\rangle$$
$$= \langle 0|(1 - N)|0\rangle = 1 \tag{9.18}$$

Thus $C = 1$. Hence

$$\hat{A}^\dagger|0\rangle = |1\rangle$$

Now

$$\hat{A}^\dagger|1\rangle = [N, \hat{A}^\dagger]|1\rangle = N\hat{A}^\dagger|1\rangle - \hat{A}^\dagger|1\rangle$$

Thus

$$N\hat{A}^\dagger|1\rangle = 2\hat{A}^\dagger|1\rangle$$

Thus $\hat{A}^\dagger|1\rangle$ is either an eigenstate of N with eigenvalue 2 or is zero. But N has only two eigenvalues 0 or 1. Hence

$$\hat{A}^\dagger|1\rangle = 0$$

Q9.7 Let $\hat{B}$ and $\hat{C}$ be two anticommutating operators

$$\hat{B}\hat{C} + \hat{C}\hat{B} = 0.$$

Let $|\psi\rangle$ be a simultaneous eigenstate of both $\hat{B}$ and $\hat{C}$. What can be said about the corresponding eigenvalues?

If $\hat{C}^2 = 1$ holds, what does it imply for the eigenvalue of $\hat{B}$?

Solution: Eigenvalue equations are

$$\hat{B}|\psi\rangle = b|\psi\rangle \qquad (9.19)$$

$$\hat{C}|\psi\rangle = c|\psi\rangle \qquad (9.20)$$

as $|\psi\rangle$ is a simultaneous eigenstate of $\hat{B}$ and $\hat{C}$. Thus from Eq. (9.19) and (9.20)

$$\hat{C}\hat{B}|\psi\rangle = bc|\psi\rangle \qquad (9.21)$$

$$\hat{B}\hat{C}|\psi\rangle = cb|\psi\rangle \qquad (9.22)$$

Thus

$$(\hat{B}\hat{C} + \hat{C}\hat{B})|\psi\rangle = 2bc|\psi\rangle \qquad (9.23)$$

Now

$$\hat{B}\hat{C} + \hat{C}\hat{B} = 0$$

$$\therefore \qquad bc|\psi\rangle = 0$$

Therefore either

$$|\psi\rangle = 0$$

$$\text{or } bc = 0$$

i.e. one of the eigenvalues is zero. If $\hat{C}^2 = 1$, then $c = \pm 1$, $b = 0$. Thus if $\hat{C}^2 = 1$, the eigenvalue of $\hat{B}$ must be zero.

Q9.8 Show that for a simple harmonic oscillator

$$\langle n|\hat{x}^2|0\rangle = \begin{cases} -\hbar\sqrt{2}m\omega, & \text{for } n = 2 \\ 0, & \text{for } n \neq 2 \end{cases} \qquad (9.24)$$

Further show that

$$\langle 0|\hat{x}^2|0\rangle = \frac{\hbar}{2m\omega}.$$

Solution:

$$\langle n|\hat{x}^2|0\rangle = \sum_m \langle n|\hat{x}|m\rangle\langle m|\hat{x}|0\rangle = -\frac{\hbar}{2m\omega}$$

$$\times \sum_m (\sqrt{m}\delta_{n,m-1} - \sqrt{m+1}\delta_{n,m+1})(-\delta_{m1}),$$

$$(9.25)$$

on using Eq. (9.16) (of problem 9.5). Hence

$$\langle n|\hat{x}^2|0\rangle = \frac{\hbar}{2m\omega}(\delta_{n0} - \sqrt{2}\delta_{n2})$$

$$= -\frac{\sqrt{2}\hbar}{2m\omega}\begin{cases} \text{for } n = 2 \\ 0 \quad n \neq 2, n \neq 0 \end{cases}$$

and

$$\langle 0|x^2|0\rangle \equiv \frac{\hbar}{2m\omega}$$

Q9.9 In the three dimensional vector space with an orthonormal set of basis vectors $\{|1\rangle, |2\rangle, |3\rangle\}$

(a) Find the matrix representation of the following operators

$$|1\rangle\langle 1| \ , \quad |2\rangle\langle 2|, \quad |1\rangle\langle 2| - |2\rangle\langle 1|, \quad |1\rangle\langle 3| - |2\rangle\langle 3| + |2\rangle\langle 1|$$

$$2|1\rangle\langle 1| - \frac{1}{\sqrt{2}}|3\rangle\langle 2| + i|2\rangle\langle 2|.$$

(b) Which of the above operators are hermitian?
(c) Let $\hat{A} = i|1\rangle\langle 2| - i|2\rangle\langle 1| + |3\rangle\langle 3|$. Show that $\hat{A}$ is hermitian and find the eingevalues and eigenstates of $\hat{A}$?
(d) Show that eigenstates of $\hat{A}$ are orthogonal.

Solution: a)

$$|1\rangle = \begin{pmatrix} 1 \\ 0 \\ 0 \end{pmatrix}, \quad |2\rangle = \begin{pmatrix} 0 \\ 1 \\ 0 \end{pmatrix}, \quad |3\rangle = \begin{pmatrix} 0 \\ 0 \\ 1 \end{pmatrix}, \qquad (9.26)$$

$$|1\rangle\langle 1| \;:\; \begin{pmatrix} 1 & 0 & 0 \\ 0 & 0 & 0 \\ 0 & 0 & 0 \end{pmatrix}$$

$$|2\rangle\langle 2| \;:\; \begin{pmatrix} 0 & 0 & 0 \\ 0 & 1 & 0 \\ 0 & 0 & 0 \end{pmatrix}$$

$$|1\rangle\langle 2| - |2\rangle\langle 1| \;:\; \begin{pmatrix} 0 & 1 & 0 \\ -1 & 0 & 0 \\ 0 & 0 & 0 \end{pmatrix}$$

$$|1\rangle\langle 3| - |2\rangle\langle 3| + |2\rangle\langle 1| \;:\; \begin{pmatrix} 0 & 0 & 1 \\ 1 & 0 & -1 \\ 0 & 0 & 0 \end{pmatrix}$$

$$2|1\rangle\langle 1| - \frac{1}{\sqrt{2}}|3\rangle\langle 2| + i|2\rangle\langle 2| \;:\; \begin{pmatrix} 2 & 0 & 0 \\ 0 & i & 0 \\ 0 & -\dfrac{1}{\sqrt{2}} & 0 \end{pmatrix}$$

b)

$$(|1\rangle\langle 1|)^{\dagger} = |1\rangle\langle 1| \qquad \text{hermitian}$$

$$(|2\rangle\langle 2|)^{\dagger} = |2\rangle\langle 2| \qquad \text{hermitian}$$

$$(|1\rangle\langle 2| - |2\rangle\langle 1|)^{\dagger} = |2\rangle\langle 1| - |1\rangle\langle 2|$$

$$\times \text{ Not hermitian}$$

$$(|1\rangle\langle 3| - |2\rangle\langle 3| + |2\rangle\langle 1|)^{\dagger} = |3\rangle\langle 1| - |3\rangle\langle 2| + |1\rangle\langle 2|$$

$$\times \text{ Not hermitian}$$

$$(2|1\rangle\langle 1| - \frac{1}{\sqrt{2}}|3\rangle\langle 2| + i|2\rangle\langle 2|)^\dagger = 2|1\rangle\langle 1| - \frac{1}{\sqrt{2}}|2\rangle\langle 3|$$

$$- i|2\rangle\langle 2|$$

Not hermitian

c)

$$\hat{A} = i|1\rangle\langle 2| - i|2\rangle\langle 1| + |3\rangle\langle 3|$$

$$\hat{A}^\dagger = -i|2\rangle\langle 1| + i|1\rangle\langle 2| + |3\rangle\langle 3| = \hat{A}$$

$$\hat{A} : \begin{pmatrix} 0 & i & 0 \\ -i & 0 & 0 \\ 0 & 0 & 1 \end{pmatrix}$$

The eigenvalues are solution of $det|\hat{A} - aI| = 0$ or

$$\begin{vmatrix} -a & i & 0 \\ -i & -a & 0 \\ 0 & 0 & 1-a \end{vmatrix} = 0$$

Thus

$$(1-a)(a^2 - 1) = 0$$

The eigenvalues are

$$a = 1, -1, 1 \tag{9.27}$$

To find eigenstates, we have

$$\begin{pmatrix} 0 & i & 0 \\ -i & 0 & 0 \\ 0 & 0 & 1 \end{pmatrix} \begin{pmatrix} x \\ y \\ z \end{pmatrix} = a \begin{pmatrix} x \\ y \\ z \end{pmatrix} \tag{9.28}$$

The eigenstates are:

$$|a = 1\rangle = \frac{1}{\sqrt{2}} \begin{pmatrix} -1 \\ i \\ 0 \end{pmatrix}, \quad |a = -1\rangle = \frac{1}{\sqrt{2}} \begin{pmatrix} 1 \\ i \\ 0 \end{pmatrix}$$

$$|a = 1\rangle = \begin{pmatrix} 0 \\ 0 \\ 1 \end{pmatrix} \tag{9.29}$$

d)

$$(0 \ \ 0 \ \ 1) \begin{pmatrix} -1 \\ i \\ 0 \end{pmatrix} = 0$$

$$(0 \ \ 0 \ \ 1) \begin{pmatrix} -1 \\ i \\ 0 \end{pmatrix} = 0, \text{ etc}$$

Q9.10 For the Hamiltonian

$$H = \frac{\hat{p}^2}{2m} + V(r)$$

$$[\hat{p}_i, H] = -i\hbar \frac{\partial}{\partial x_i} V(r).$$

Using the above result, show that

$$[\hat{p}_j, [\hat{p}_i, H]] = (-i\hbar)^2 \frac{\partial}{\partial x_j} \frac{\partial}{\partial x_i} V(r). \qquad (9.30)$$

From the relation:

$$\langle m|[\hat{p}_j, [\hat{p}_i, H]]|m \rangle = (-i\hbar)^2 \langle m| \left(\frac{\partial}{\partial x_j} \frac{\partial}{\partial x_i} V(r) \right) |m \rangle,$$

derive the sum rule

$$\sum_n (E_m - E_n)\langle m|\hat{p}_i|n\rangle\langle n|\hat{p}_i|m\rangle = -\frac{\hbar^2}{2}\langle m| \nabla^2 V|m\rangle,$$

i.e.

$$\sum_n (E_n - E_m)|(\mathbf{p})_{mn}|^2 = \frac{\hbar^2}{2} \int |\psi_m(\mathbf{x})|^2 \nabla^2 V d^3x$$

$$= \frac{\hbar^2}{2} \ \ 4\pi e^2 Z |\psi_m(0)|^2$$

where for the Coulomb potential

$$\nabla^2 V = 4\pi e^2 Z \delta^3(\vec{x})$$

Solution:

$$[\hat{p}_j, [\hat{p}_i, H]] = (-i\hbar)\left[\hat{p}_j, \frac{\partial V}{\partial x_i}\right]$$

$$= (-i\hbar)^2 \frac{\partial}{\partial x_j}\frac{\partial}{\partial x_i}V(r) \qquad (9.31)$$

Thus

$$\langle m|[\hat{p}_j, [\hat{p}_i, H]]|m\rangle = -\hbar^2 \left\langle \frac{\partial}{\partial x_j}\frac{\partial}{\partial x_i}V\right\rangle \qquad (9.32)$$

Now

$$\text{L.H.S} = \sum_n \left[\langle m|\hat{p}_j|n\rangle\langle n|[\hat{p}_i, H]|m\rangle\right.$$

$$\left. - \langle m|[\hat{p}_i, H]|n\rangle\langle n|\hat{p}_j|m\rangle\right]$$

$$= \sum_n [\langle m|\hat{p}_j|n\rangle(E_m - E_n)\langle n|\hat{p}_i|m\rangle$$

$$- (E_n - E_m)\langle m|\hat{p}_i|n\rangle\langle n|\hat{p}_j|m\rangle]$$

Multiply both sides of Eq. (9.32) by δ_{ij} and sum over j. Then

$$\sum_n (E_n - E_m)\langle m|\hat{p}_i|n\rangle\langle n|\hat{p}_i|m\rangle$$

$$= \frac{\hbar^2}{2}\int\int d^3x d^3x' \langle m|\mathbf{x}\rangle\langle \mathbf{x}|\nabla^2 V|\mathbf{x}'\rangle\langle \mathbf{x}'|m\rangle\rangle$$

$$= \frac{\hbar^2}{2}\int d^3x d^3x' \psi_m^*(\mathbf{x})(\nabla^2 V)\delta(\mathbf{x} - \mathbf{x}')\psi_m(\mathbf{x}')$$

$$= \frac{\hbar^2}{2}\int |\psi_m(\mathbf{x})|^2 \nabla^2 V d^3x$$

$$= \frac{\hbar^2}{2}4\pi Z e^2 |\psi_m(0)|^2$$

where we have used

$$\nabla^2 V = 4\pi Z e^2 \delta^3(x)$$

Q9.11 Show that 3×3 matrices

$$(G_i)_{jk} = -i\hbar\epsilon_{ijk}$$

satisfy the relation

$$(G_iG_j - G_jG_i)_{mk} = -i\hbar\epsilon_{ijn}(G_n)_{mk}$$

i.e.

$$[G_i, G_j] = i\hbar\epsilon_{ijk}G_n,$$

showing that matrices G's satisfy the commutation relation of angular momentum and as such give the adjoint representation of group O_3 and represent spin 1. Write these matrices explicitly:

$$G_1 = \hbar \begin{pmatrix} 0 & 0 & 0 \\ 0 & 0 & -i \\ 0 & -i & 0 \end{pmatrix}, \quad G_2 = \hbar \begin{pmatrix} 0 & 0 & i \\ 0 & 0 & 0 \\ -i & 0 & 0 \end{pmatrix}$$

$$G_3 = \hbar \begin{pmatrix} 0 & -i & 0 \\ i & 0 & 0 \\ 0 & 0 & 0 \end{pmatrix}$$

Show that the eigenvalues of G_3 are $\hbar, 0, -\hbar$.
In the basis, in which G_3 is diagonal the corresponding eigenvectors $|x\rangle$ are:

$$|1\rangle = \begin{pmatrix} 1 \\ 0 \\ 0 \end{pmatrix}, |0\rangle = \begin{pmatrix} 0 \\ 1 \\ 0 \end{pmatrix}, |-1\rangle = \begin{pmatrix} 0 \\ 0 \\ 1 \end{pmatrix}.$$

Show that the eigenvectors, $|y\rangle$ in the basis in which G_3 is not diagonal are:

$$|G_3 = +1\rangle = \frac{1}{\sqrt{2}} \begin{pmatrix} -1 \\ -i \\ 0 \end{pmatrix}, |G_3 = 0\rangle = \begin{pmatrix} 0 \\ 0 \\ 1 \end{pmatrix},$$

$$|G_3 = -1\rangle = \frac{1}{\sqrt{2}} \begin{pmatrix} -1 \\ i \\ 0 \end{pmatrix}.$$

Show that the matrix

$$U_{mi} = \langle y_m | x_i \rangle.$$

which connect the two bases is given by

$$U = \frac{1}{\sqrt{2}} \begin{pmatrix} -1 & i & 0 \\ 0 & 0 & \sqrt{2} \\ -1 & -i & 0 \end{pmatrix}.$$

Solution:

$$(G_i)_{jk} = -i\hbar \epsilon_{ijk}, \tag{9.33}$$

$$(G_i)_{mn}(G_j)_{nk} = (-i\hbar)^2 \epsilon_{imn} \epsilon_{jnk}$$
$$= \hbar^2 (\delta_{ij}\delta_{mk} - \delta_{ik}\delta_{mj})$$

$$(G_j)_{mn}(G_i)_{nk} = (-i\hbar)^2 \epsilon_{jmn} \epsilon_{ink}$$
$$= \hbar^2 (\delta_{ji}\delta_{mk} - \delta_{jk}\delta_{mi})$$

Thus

$$(G_i G_j - G_j G_i)_{mk} = \hbar^2 [\delta_{jk}\delta_{im} - \delta_{ik}\delta_{jm}]$$
$$= -\hbar^2 (\epsilon_{kmn} \epsilon_{ijn})$$
$$= i\hbar(-i\hbar)\epsilon_{ijn} \epsilon_{nmk} = i\hbar \epsilon_{ijn} (G_n)_{mk}$$

Hence

$$[G_i, G_j] = i\hbar \epsilon_{ijn} G_n$$

Matrices G_i's satisfy the C.R. of Angular momentum. Thus 3×3 matrices represent the spin 1. From Eq. (9.33)

$$G_1 = = \hbar \begin{pmatrix} 0 & 0 & 0 \\ 0 & 0 & -i \\ 0 & i & 0 \end{pmatrix}$$

Similarly,

$$G_2 = = \hbar \begin{pmatrix} 0 & 0 & i \\ 0 & 0 & -i \\ -i & 0 & 0 \end{pmatrix}$$

$$G_3 = = \hbar \begin{pmatrix} 0 & -i & 0 \\ i & 0 & 0 \\ 0 & 0 & 0 \end{pmatrix} \tag{9.34}$$

In the basis, in which G_3 is a diagonal matrix, the eigenstates $|x_i\rangle$ of G_3 are

$$|x_1\rangle = \begin{pmatrix} 1 \\ 0 \\ 0 \end{pmatrix}, \quad |x_2\rangle = \begin{pmatrix} 0 \\ 1 \\ 0 \end{pmatrix}, \quad |x_3\rangle = \begin{pmatrix} 0 \\ 0 \\ 1 \end{pmatrix}, \tag{9.35}$$

corresponding to the eigenvalues $1, 0, -1$ respectively. In the basis in which G_3 is given by Eq. (9.34), the eigenstates $|y_m\rangle$ of G_3 are

$$|y_1\rangle = \frac{1}{\sqrt{2}} \begin{pmatrix} -1 \\ -i \\ 0 \end{pmatrix}, \quad |y_2\rangle = \begin{pmatrix} 0 \\ 1 \\ 0 \end{pmatrix}, \quad |y_3\rangle = \frac{1}{\sqrt{2}} \begin{pmatrix} 0 \\ 0 \\ 1 \end{pmatrix} \tag{9.36}$$

Corresponding to eigenvalues $+1, 0, -1$ of G_3 given in Eq. (9.34). We note

$$|x_i\rangle\langle x_i| = 1, \quad |y_m\rangle\langle y_m| = 1$$
$$|x_i\rangle = U_{ni}|y_n\rangle$$
$$\langle y_m|x_i\rangle = U_{ni}\delta_{mn}$$
$$= U_{mi}$$

Thus

$$|x_i\rangle = \sum_n |y_n\rangle\langle y_n|x_i\rangle$$

$$= \sum_n U_{ni}|y_n\rangle$$

$$\langle y_m|x_i\rangle = \sum_n U_{ni}\delta_{mn} = U_{mi} \tag{9.37}$$

Using Eqs. (9.35) and (9.36), from Eq. (9.37), we obtain

$$U = \frac{1}{\sqrt{2}} \begin{pmatrix} -1 & i & 0 \\ 0 & 0 & \sqrt{2} \\ -1 & -i & 0 \end{pmatrix}$$

Additional Problems:

Q1 (i) Show that

$$e^{\widehat{A}} e^{\widehat{B}} = e^{\widehat{A}+\widehat{B}+\frac{1}{2}[\widehat{A},\widehat{B}]}$$

if $\widehat{A}$ and $\widehat{B}$ commute with $[\widehat{A}, \widehat{B}]$

(ii)

$$\langle 0|e^{ik\widehat{x}}|0\rangle = e^{-k^2 \frac{\langle 0|\widehat{x}^2|0\rangle}{2}}$$

for one dimensional simple harmonic oscillator where $|0\rangle$ is the ground state of the oscillator.

Solution:

(i) Consider

$$\widehat{F}(\lambda) = e^{\lambda\widehat{A}} e^{\lambda\widehat{B}}$$

$$\frac{d\widehat{F}(\lambda)}{d\lambda} = \widehat{A}\widehat{F}(\lambda) + e^{\lambda\widehat{A}}\widehat{B}e^{\lambda\widehat{B}}$$

$$= \widehat{A}\widehat{F}(\lambda) + e^{\lambda\widehat{A}}\widehat{B}e^{\lambda\widehat{A}}e^{-\lambda\widehat{A}}e^{\lambda\widehat{B}}$$

$$= \widehat{A}\widehat{F}(\lambda) + [\widehat{B} + \lambda[\widehat{A}, \widehat{B}]]\widehat{F}(\lambda)$$

where we have used Problem 10.3 (see next chapter) with $[\widehat{A}, [\widehat{A}, \widehat{B}]] = 0$ etc.

Thus

$$\frac{d\widehat{F}(\lambda)}{d\lambda} = (\widehat{A} + \widehat{B})\widehat{F}(\lambda) + \lambda[\widehat{A}, \widehat{B}]\widehat{F}(\lambda)$$

Under the condition of the problem $\widehat{F}(\lambda)$ commutes with $[\widehat{A}, \widehat{B}]$ and also with $(\widehat{A} + \widehat{B})$. Thus the intergation give

$$\widehat{F}(\lambda) = e^{\lambda(\widehat{A}+\widehat{B}+\frac{\lambda^2}{2}[\widehat{A},\widehat{B}])}$$

Putting $\lambda = 1$, we get

$$e^{\widehat{A}} e^{\widehat{B}} = e^{(\widehat{A}+\widehat{B})+\frac{1}{2}[\widehat{A},\widehat{B}]}$$

(ii) For one dimensional oscillator

$$a^\dagger - a = ik\widehat{x}, \quad k = \sqrt{\frac{2mw}{\hbar}}$$

In part (i), take

$$\widehat{A} = a^\dagger, \quad \widehat{B} = -a, \quad [\widehat{A}, \widehat{B}] = [a, a^\dagger]$$

Then

$$e^{a^\dagger} e^{-a} = e^{ik\widehat{x} + \frac{1}{2}}$$

Thus

$$\langle 0 | e^{a^\dagger} e^{-a} | 0 \rangle = \langle 0 | e^{ik\widehat{x}} | 0 \rangle e^{\frac{1}{2}}$$

But

$$e^{-a} | 0 \rangle = (1 - a + \cdots) | 0 \rangle = | 0 \rangle$$

Similarly

$$\langle 0 | e^{a^\dagger} = \langle 0 |$$

$$\langle 0 | e^{ik\widehat{x}} | 0 \rangle = e^{-\frac{1}{2}}$$

Now from Problem 9.8, $\langle 0 | \widehat{x}^2 | 0 \rangle = \frac{\hbar}{2mw} = \frac{1}{k^2}$.
Thus we have

$$\langle 0 | e^{ik\widehat{x}} | 0 \rangle = e^{-\frac{k^2}{2} \langle 0 | \widehat{x}^2 | 0 \rangle}$$

Chapter 10

Heisenberg Equation of Motion, Invariance Principle and Path Integral

Q10.1 Show that for the simple harmonic oscillator, the Heisenberg equation of motion gives

$$i\hbar\dot{a}(t) = [a(t), H] = \hbar\omega a(t)$$

which has the solution

$$a(t) = a(0)e^{-i\omega t} = ae^{-i\omega t}$$
$$a^\dagger(t) = a^\dagger e^{i\omega t}.$$

Hence show that

$$\hat{q}(t) = \hat{q}(0)\cos\omega t + \frac{\hat{p}(0)}{m\omega}\sin\omega t$$
$$\hat{p}(t) = [\hat{p}(0)\cos\omega t - m\omega\hat{q}(0)\sin\omega t].$$

$$[\hat{q}(t_1), \hat{q}(t_2)] = \frac{i\hbar}{m\omega}\sin\omega(t_2 - t_1) \neq 0, \quad \text{if } t_2 \neq t_1$$

$$[\hat{q}(t_2), \hat{p}(t_1)] = i\hbar\cos\omega(t_2 - t_1) \neq i\hbar. \quad \text{if } t_2 \neq t_1$$

Thus canonical quantum conditions hold only for equal time $t_2 = t_1 = t$

Solution: For harmonic oscillator

$$H = \hbar\omega \left(a^\dagger a + \frac{1}{2} \right)$$

The time dependent annihilation operator $a(t)$ is defined as

$$a(t) = e^{iHt/\hbar} a(0) e^{-iHt/\hbar} \qquad (10.1)$$

The time derivative of Eq. (10.1) gives

$$\frac{d}{dt} a(t) = \frac{d}{dt} [e^{iHt/\hbar} a(0) e^{-iHt/\hbar}]$$

$$= -\frac{i}{\hbar}(a(t)H - Ha(t))$$

$$\frac{d}{dt} a(t) = -\frac{i}{\hbar}[a(t), H]$$

For the simple harmonic oscillator

$$\hat{q} = \sqrt{\frac{\hbar}{2m\omega}} \frac{1}{i}(a^\dagger - a)$$

$$\hat{p} = \sqrt{\frac{m\hbar\omega}{2}}(a^\dagger + a)$$

Thus

$$\hat{q}(t) = (-i)\sqrt{\frac{\hbar}{2m\omega}} [a^\dagger(0)e^{i\omega t} - a(0)e^{-i\omega t}]$$

$$= \sqrt{\frac{\hbar}{2m\omega}} [-i\cos\omega t(a^\dagger(0) - a(0))$$

$$+ \sin\omega t(a^\dagger(0) + a(0))]$$

$$= \left[\hat{q}(0)\cos\omega t + \frac{\hat{p}(0)}{m\omega}\sin\omega t \right]$$

$$\hat{p}(t) = \sqrt{\frac{m\hbar\omega}{2}} [a^\dagger(0)e^{i\omega t} + a(0)e^{-i\omega t}]$$

$$= [\hat{p}(0)\cos\omega t - m\omega\hat{q}(0)\sin\omega t]$$

Hence

$$
\begin{aligned}
[\hat{q}(t_1), \hat{q}(t_2)] &= \left[\hat{q}(0) \cos \omega t_1 + \frac{\hat{p}(0)}{m\omega} \sin \omega t_1, \hat{q}(0) \cos \omega t_2 \right. \\
&\qquad \left. + \frac{\hat{p}(0)}{m\omega} \sin \omega t_2 \right] \\
&= \frac{1}{m\omega} \cos \omega t_1 \sin \omega t_2 [\hat{q}(0), \hat{p}(0)] \\
&\quad + \frac{1}{m\omega} \sin \omega t_1 \cos \omega t_2 [\hat{p}(0), \hat{q}(0)] \\
&= \frac{i\hbar}{m\omega} \sin \omega (t_2 - t_1) \neq 0
\end{aligned}
$$

for $t_2 \neq t_1$

Similarly

$$
\begin{aligned}
[\hat{q}(t_2), \hat{p}(t_1)] &= \cos \omega t_2 \cos \omega t_1 [\hat{q}(0), \hat{p}(0)] \\
&\quad - \sin \omega t_2 \sin \omega t_1 [\hat{p}(0), \hat{q}(0)] \\
&= i\hbar \cos \omega (t_2 - t_1) \neq 0
\end{aligned}
$$

for $t_2 \neq t_1$.

Q10.2 Show that for the angular momentum $\boldsymbol{L} = \boldsymbol{x} \times \boldsymbol{p}$, $L_i = \epsilon_{iln} x_l p_n$, the classical PB is

$$
\begin{aligned}
(L_i, L_j)_{PB} &= \epsilon_{iln} \epsilon_{jrs} \sum_k (\delta_{lk} p_n \delta_{sk} x_r - \delta_{nk} x_l \delta_{rk} p_s) \\
&= x_i p_j - x_j p_i \\
&= \epsilon_{ijk} L_k
\end{aligned}
$$

Thus in quantum mechanics

$$
\frac{1}{i\hbar} \left[\hat{L}_i, \hat{L}_j \right] = \epsilon_{ijk} \hat{L}_k
$$

$$
\left[\hat{L}_i, \hat{L}_j \right] = i\hbar \epsilon_{ijk} \hat{L}_k
$$

Solution:

$$L_i = \epsilon_{iln} x_l p_n \qquad (10.2)$$

$$L_j = \epsilon_{jl'n'} x_{l'} p_{n'} \qquad (10.3)$$

Using the definition of Poisson's bracket (cf. Eq. (10.30) from the text) for L_i and L_j

$$(L_i, L_j)_{PB} = \left(\frac{\partial L_i}{\partial x_k} \frac{\partial L_j}{\partial p_k} - \frac{\partial L_i}{\partial p_k} \frac{\partial L_j}{\partial x_k} \right) \qquad (10.4)$$

Using Eqs. (10.2) and (10.3) in Eq. (10.4) gives

$$
\begin{aligned}
(L_i, L_j)_{PB} &= \epsilon_{iln} \epsilon_{jl'n'} \left(p_n \delta_{lk} x_{l'} \delta_{kn'} - x_l \delta_{kn} p_{n'} \delta_{l'k} \right) \\
&= \left(\epsilon_{ikn} \epsilon_{jl'k} p_n x_{l'} - \epsilon_{ilk} \epsilon_{jkn'} x_l p_{n'} \right) \\
&= -\delta_{ij} p_n x_n + p_j x_i + \delta_{ij} x_l p_l - x_j p_i \\
&= (x_i p_j - x_j p_i) \\
&= \epsilon_{ijk} L_k
\end{aligned}
$$

Thus in QM

$$\frac{1}{i\hbar} [\hat{L}_i, \hat{L}_j] = \epsilon_{ijk} \hat{L}_k$$

$$[\hat{L}_i, \hat{L}_j] = i\hbar \epsilon_{ijk} \hat{L}_k$$

Q10.3 Show that

$$e^{\hat{A}} \hat{B} e^{-\hat{A}} = \hat{B} + [\hat{A}, \hat{B}] + \frac{1}{2} [\hat{A}, [\hat{A}, \hat{B}]] + \cdots .$$

Hint: Consider $\hat{f}(\lambda) = e^{\lambda \hat{A}} \hat{B} e^{-\lambda \hat{A}}$,
and show that

$$\frac{d\hat{f}(\lambda)}{d\lambda} = [\hat{A}, \hat{f}(\lambda)]$$

$$\frac{d^2 \hat{f}(\lambda)}{d\lambda^2} = [\hat{A}, [\hat{A}, \hat{f}(\lambda)]] \quad \text{etc.}$$

and use the Taylor's series

$$\hat{f}(\lambda) = \hat{f}(0) + \frac{\lambda}{1!} \frac{d\hat{f}(\lambda)}{d\lambda!}\bigg|_{\lambda=0} + \frac{\lambda^2}{2!} \frac{d^2 f}{d\lambda^2}\bigg|_{\lambda=0} + \cdots .$$

to get the required quantity $\hat{f}(\lambda)$

Solution: Consider

$$\hat{f}(\lambda) = e^{\lambda\hat{A}} \hat{B} e^{-\lambda\hat{A}} \tag{10.5}$$

Multiplying Eq. (10.5) with $e^{-\lambda\hat{A}}$ from the left and differentiating it w.r.t λ gives

$$(-\hat{A})e^{-\lambda\hat{A}}\hat{f}(\lambda) + e^{-\lambda\hat{A}}\frac{d\hat{f}(\lambda)}{d\lambda} = \hat{B}(-\hat{A})e^{-\lambda\hat{A}} \tag{10.6}$$

Multiplying both sides of Eq. (10.6) by $e^{\lambda\hat{A}}$ on the left gives

$$\frac{d\hat{f}(\lambda)}{d\lambda} = \hat{A}\hat{f}(\lambda) - \hat{f}(\lambda)\hat{A} \tag{10.7}$$

$$= [\hat{A}, \hat{f}(\lambda)]$$

Differentiating Eq. (10.7) gives

$$\frac{d^2\hat{f}(\lambda)}{d\lambda^2} = \left[\hat{A}, \frac{d\hat{f}(\lambda)}{d\lambda}\right]$$

$$\frac{d^2\hat{f}(\lambda)}{d\lambda^2} = \left[\hat{A}, [\hat{A}, \hat{f}(\lambda)]\right] \tag{10.8}$$

Making use of the Taylor series in Eqs. (10.7) and (10.8) gives (c.f. $\hat{f}(0) = \hat{B}$)

$$\hat{f}(\lambda) = \hat{B} + \frac{\lambda}{1!}[\hat{A}, \hat{B}] + \frac{\lambda^2}{2!}\left[\hat{A}, [\hat{A}, \hat{B}]\right] + \cdots . \tag{10.9}$$

Using $\lambda = 1$ in Eq. (10.9) gives

$$\hat{f}(1) = e^{\hat{A}} \hat{B} e^{-\hat{A}}$$

$$= \hat{B} + [\hat{A}, \hat{B}] + \frac{1}{2!}\left[\hat{A}, [\hat{A}, \hat{B}]\right] + \cdots$$

Q10.4 Let x and p_x be the coordinate and linear momentum in one dimension. Evaluate the classical PB

$$(x, F(p_x))_{PB}$$

Let $\hat{x}$ and $\hat{p}_x$ be corresponding quantum mechanical operators. Then using the rule to go from classical mechanics to quantum mechanics, evaluate the commutator.

$$\left[\hat{x}, \exp\left(\frac{ia\hat{p}_x}{\hbar}\right)\right]$$

where a is a real number. Using the above result, prove that $\exp(\frac{ia\hat{p}_x}{\hbar})|\hat{x}\rangle$ is an eigenstate of the coordinate operator $\hat{x}$ [remember $\hat{x}|\hat{x}\rangle = \hat{x}|\hat{x}\rangle$]. What is the corresponding eigenvalue. Thus $\exp(\frac{ia\hat{p}_x}{\hbar})$ gives the unitary operator corresponding to the translation.

Solution: Using the definition of Poisson's bracket

$$(x, F(p_x)) = \left(\frac{\partial x}{\partial x}\frac{\partial F(p_x)}{\partial p_x} - \frac{\partial x}{\partial p_x}\frac{\partial F(p_x)}{\partial x}\right)$$

$$= \frac{\partial F(p_x)}{\partial p_x}$$

Take

$$F(p_x) = e^{iap_x}$$

$$\frac{\partial F(p_x)}{\partial p_x} = iae^{iap_x}$$

Thus

$$(x, e^{iap_x})_{PB} = iae^{iap_x}$$

To go over to QM

$$x \to \hat{x}$$

$$p_x \to \hat{p}_x$$

$$(x, e^{iap_x})_{PB} \to \frac{1}{i\hbar}[\hat{x}, e^{ia\hat{p}_x}]$$

Thus

$$[\hat{x}, e^{ia\hat{p}_x}] = -a\hbar e^{ia\hat{p}_x}$$

Replace a by $a/\hbar$ in above equation

$$[\hat{x}, e^{ia\hat{p}_x/\hbar}] = -a e^{ia\hat{p}_x/\hbar}$$

Hence

$$\hat{x} e^{ia\hat{p}_x/\hbar}|x'\rangle = -a e^{ia\hat{p}_x/\hbar}|x'\rangle + e^{ia\hat{p}_x/\hbar}\hat{x}|x'\rangle$$
$$= (x' - a)e^{ia\hat{p}_x/\hbar}|x'\rangle$$

Thus if $e^{ia\hat{p}_x/\hbar}|x'\rangle$ is not zero, it is an eigenstate of $\hat{x}$ with eigenvalue $(x' - a)$.

Q10.5 Consider the Hamiltonian describing a free particle in an external field ϵ (constant in time)

$$H = \frac{\hat{p}^2}{2m} - e\epsilon\hat{x}.$$

Calculate the operators $\hat{x}(t)$ and $\hat{p}(t)$ in the Heisenberg picture.

Suppose at $t = 0$, the particle is in the state $|\psi_0\rangle$ whose wave function in the x-representation is $\psi_0(x) = \langle x|\psi_0\rangle = e^{ikx}\phi(x)$, where $\phi(-x) = \phi(x)$ and $\phi(x)$ is real and $\int |\phi(x)|^2 dx = 1$. Define the uncertainty $(\Delta A)_t$ in the observable $\hat{A}$ at time t as

$$(\Delta A)_t = \{\langle(\hat{A}(t) - \langle\hat{A}(t)\rangle)^2\rangle\}^{\frac{1}{2}},$$

where $\langle\hat{A}(t)\rangle = \langle\psi_0|\hat{A}(t)|\psi_0\rangle$ in the Heisenberg picture. Then prove the following:

(i)

$$\langle\hat{x}(0)\rangle = 0, \quad \langle\hat{p}(0)\rangle = \hbar k, \quad \langle\hat{p}(0)^2\rangle = 2m\langle E\rangle$$
$$\langle\hat{x}(0)\hat{p}(0) + \hat{p}(0)\hat{x}(0)\rangle = 0.$$

(ii)

$$(\Delta x)_t = \left\{ (\Delta x)_0^2 + \frac{2t^2}{m} \left(\langle E \rangle - \frac{\hbar^2 k^2}{2m} \right) \right\}^{\frac{1}{2}}$$

$$(\Delta p)_t = \left\{ 2m \left(\langle E \rangle - \frac{\hbar^2 k^2}{2m} \right) \right\}^{\frac{1}{2}}$$

$$= (\Delta p)_0$$

Solution: To calculate $\hat{x}(t)$ and $\hat{p}(t)$ in the Heisnberg picture we use the following

$$i\hbar \frac{d\hat{p}(t)}{dt} = [\hat{p}(t), H]$$

$$= -e\epsilon[\hat{p}(t), \hat{x}(t)]$$

$$= i\hbar e\epsilon$$

$$\therefore \hat{p}(t) = e\epsilon t + \hat{p}(0)$$

Now

$$i\hbar \frac{d\hat{x}}{dt} = [\hat{x}(t), H]$$

$$= \frac{1}{2m} [\hat{x}(t), \hat{p}^2(t)]$$

$$= \frac{1}{2m} 2i\hbar \hat{p}(t)$$

$$\frac{d\hat{x}}{dt} = \frac{\hat{p}(t)}{m}$$

$$\hat{x}(t) = \hat{x}(0) + \frac{\hat{p}(0)}{m} t + \frac{e}{2m} \epsilon t^2$$

Put $\hat{p}(0) = \hat{p}$, $\hat{x}(0) = \hat{x}$

$$\hat{p}(t) = \hat{p} + e\epsilon t$$

$$\hat{x}(t) = \hat{x} + \frac{\hat{p}}{m} t + \frac{e}{2m} \epsilon t^2$$

Define the uncertainty for an observable $\hat{A}$ as

$$\langle \Delta A \rangle_t = \sqrt{\langle (\hat{A}(t) - \langle \hat{A}(t) \rangle)^2 \rangle}$$

At $t = 0$, the system is in the state $|\psi_0\rangle$, where the wave function in x-representation is $\psi_0(x) = e^{ikx}\phi(x)$. $\phi(-x) = \phi(x)$ is real and $\int |\phi(x)|^2 dx = 1$

$$\langle \hat{x}(t) \rangle = \langle \psi_0 | \hat{x}(t) | \psi_0 \rangle$$

$$= \langle \hat{x}(0) \rangle + \frac{t}{m} \langle \hat{p}(0) \rangle + \frac{e\epsilon}{m} t^2$$

$$\langle \hat{p}(t) \rangle = \langle \hat{p}(0) \rangle + e\epsilon t$$

$$\langle \hat{x}(0) \rangle = \int_{-\infty}^{+\infty} dx e^{-ikx} \phi(x) x e^{ikx} \phi(x)$$

Change $x \to -x$, the integral changes sign, giving $\langle \hat{x}(0) \rangle = 0$

$$\langle \hat{p}(0) \rangle = \int_{-\infty}^{+\infty} dx e^{-ikx} \phi(x) \left(-i\hbar \frac{\partial}{\partial x} \right) e^{ikx} \phi(x)$$

$$= \int_{-\infty}^{+\infty} dx e^{-ikx} \phi(x) \hbar k e^{ikx} \phi(x)$$

$$- i\hbar \int_{-\infty}^{+\infty} dx e^{-ikx} \phi(x) e^{ikx} \frac{\partial \phi}{\partial x} \tag{10.10}$$

The second term in Eq. (10.10) changes sign as $x \to -x$, so it vanishes, giving $\langle \hat{p}(0) \rangle = \hbar k$. Now

$$\left\langle \frac{\hat{p}^2(0)}{2m} \right\rangle = \langle H \rangle + e\epsilon \langle \hat{x}(0) \rangle \tag{10.11}$$

As we worked out earlier $\langle \hat{x}(0) \rangle = 0$, so Eq. (10.11) becomes

$$\left\langle \frac{\hat{p}^2(0)}{2m} \right\rangle = \langle E \rangle$$

$$\langle \hat{p}^2(0) \rangle = 2m \langle E \rangle$$

Now

$$\langle \hat{x}(0)\hat{p}(0) + \hat{p}(0)\hat{x}(0) \rangle$$

$$= \langle -i\hbar + 2\hat{x}(0)\hat{p}(0) \rangle$$

$$= -i\hbar + 2 \int_{-\infty}^{\infty} dx e^{-ikx} \phi(x) x \left(-i\hbar \frac{\partial}{\partial x} \right) e^{ikx} \phi(x)$$

$$= -i\hbar - 2i\hbar \left(\int_{-\infty}^{\infty} dx \phi(x) x (ik) \phi(x) \right.$$

$$+ \left. \int_{-\infty}^{\infty} dx x \phi(x) \frac{\partial \phi(x)}{\partial x} \right)$$

$$= -i\hbar - 2i\hbar \int_{-\infty}^{\infty} dx x \phi(x) \frac{\partial \phi(x)}{\partial x}$$

$$= -i\hbar - i\hbar \int_{-\infty}^{\infty} dx x \frac{\partial}{\partial x} (\phi^2)$$

$$= -i\hbar - i\hbar \left(\int_{-\infty}^{\infty} dx \frac{\partial}{\partial x} (x\phi^2) - \int_{-\infty}^{\infty} dx \phi^2 \right)$$

$$= -i\hbar + i\hbar = 0$$

where the first integral, being total divergence, can be put equl to zero, while the second integral gives 1.
Also

$$(\Delta x)_t = [\langle (\hat{x}(t) - \langle \hat{x}(t) \rangle)^2 \rangle]^{1/2}$$

$$= \left[\langle (\hat{x}(0) + \frac{\hat{p}(0)t}{m} + \frac{e}{m}\epsilon t^2 - \frac{\hbar k}{m}t - \frac{e}{m}\epsilon t^2)^2 \rangle \right]^{1/2}$$

$$= \left[\langle \hat{x}^2(0) \rangle + \frac{\langle \hat{p}^2(0) \rangle}{m}t^2 + (\hat{x}(0)\hat{p}(0) + \hat{p}(0)\hat{x}(0)) \frac{t}{m} \right.$$

$$\left. - 2\frac{\hbar k}{m}t (\langle \hat{x}(0) \rangle + \frac{\langle \hat{p}(0) \rangle}{m}t) + \frac{\hbar^2 k^2}{m^2}t^2 \right]^{1/2}$$

$$= \left[\langle x \rangle_0^2 + 2\frac{t^2}{m} \left(\Delta E - \frac{\hbar^2 k^2}{m} \right) \right]^{1/2}$$

And

$$(\Delta k)_t = [\langle(\hat{p}(t) - \langle\hat{p}(t)\rangle^2)\rangle]^{1/2}$$

$$= [\langle\hat{p}(0) + e\epsilon t - \langle\hat{p}(0)\rangle^2]^{1/2}$$

$$= [\langle(\hat{p}(0) - \langle\hat{p}(0)\rangle)^2\rangle]$$

$$= (2mE - 2\hbar^2 k^2 + \hbar^2 k^2))^{1/2}$$

$$= \left[2m(\langle E\rangle - \frac{\hbar^2 k^2}{2m})\right]^{1/2}$$

$$= (\Delta k)_0$$

Q10.6 For the Hamiltonian

$$H = \frac{1}{2}\hat{p}^2 V(\hat{q}),$$

show that the transition amplitude is given by

$$\langle q_F, t_F | q_I, t_I\rangle = \prod_{i=0}^{N-1} \left(\frac{-\iota}{2\pi\delta t}\right)^{\frac{1}{2}} \int dq_i V^{-\frac{1}{2}}(q_i) e^{i\int_{t_I}^{t_F} dtL}$$

where

$$L = \frac{1}{2}V^{-1}\dot{q}^2$$

is the Lagrangian.

Solution: To find the transition amplitude $\langle q_F, t_F | q_I, t_I\rangle$ use the equation below Eq. (10.128) from the text and put $m \to V^{-1}$

$$I_j = \frac{1}{2\pi\hbar}\left(\frac{2\pi\hbar}{i\delta t}V^{-1}(q_j)\right)^{1/2}.e^{\frac{i}{\hbar}\delta t[\frac{1}{2}V^{-1}(q_j)\left(\frac{q_{j+1}-q_j}{\delta t}\right)]}$$

Thus

$$\mathcal{K} = \prod_{i=0}^{N-1}\left(\frac{-i}{2\pi\hbar\delta t}\right)^{1/2}\int dq_i V^{-1/2}(q_i)e^{i/\hbar\int dtV^{-1}(q_i)\dot{q}^2}$$

or

$$K = \prod_{i=0}^{N-1} \left(\frac{-i}{2\pi\hbar\delta t}\right)^{1/2} \int dq_i V^{-1/2}(q_i) e^{i/\hbar \int_{t_I}^{t_F} dt L}$$

where $L = \frac{1}{2}V^{-1}\dot{q}^2$

Q10.7 If $T < t_1, t_2 < 0$, $S = \int_0^T L dt$, and

$$\int Dq \quad q(t_1)q(t_2)e^{\iota S} = \langle q_T|T(\hat{q}(t_1)\hat{q}(t_2))|q_0\rangle$$

where $q_T = q(T), q_0 = q(0)$, show that for the Harmonic oscillator

$$\langle q_T|T(\hat{q}(t_1)\hat{q}(t_2))|q_0\rangle$$
$$= \left\{ \frac{\iota\hbar}{m\omega\sin\omega t}[\theta(t_1 - t_2)\sin\omega(T - t_1)\right.$$
$$\times \sin\omega t_2 + \theta(t_2 - t_1)\sin\omega(T - t_2)\sin\omega t_1]$$
$$+ \frac{1}{\sin^2\omega T}[q_0\sin\omega(T - t_1) + q_T\sin\omega t_1]$$
$$\left. \times [q_0\sin\omega(T - t_2) + q_T\sin\omega t_2] \right\}\langle q_F|q_I\rangle$$

where

$$\langle q_F|q_I\rangle = \int Dq \ e^{\iota S}$$

with $S = \int_0^T dt[\frac{1}{2}m\dot{q}^2 - \frac{1}{2}m\omega^2 q^2]$. Calculate S_{cl} for this case. [Hint: First show that for S.H.O, $\hat{q}(t) = \frac{1}{\sin\omega T}$ $[\hat{q}_T\sin\omega t + \hat{q}_0\sin\omega(T - t)]$

Solution: From the solution of problem 10.1, we have for SHO

$$\hat{q}(t) = \hat{q}(0)\cos\omega t + \frac{\hat{p}(0)}{m\omega}\sin\omega t$$

$$\hat{q}(T) = \hat{q}(0)\cos\omega T + \frac{\hat{p}(0)}{m\omega}\sin\omega T$$

Eliminating $\hat{p}(0)$, we have

$$\hat{q}(t) = \frac{1}{\sin \omega T} [\hat{q}_T \sin \omega t + \hat{q}(0) \sin \omega (T - t)]$$

To find $\langle q_T | T(\hat{q}(t_1)\hat{q}(t_2)) | q_0 \rangle$ we use

$$\langle q_T | T(\hat{q}(t_1)\hat{q}(t_2)) | q_0 \rangle$$
$$= \theta(t_1 - t_2)\langle q_T | (\hat{q}(t_1)\hat{q}(t_2)) | q_0 \rangle$$
$$+ \theta(t_2 - t_1)\langle q_T | (\hat{q}(t_2)\hat{q}(t_1)) | q_0 \rangle$$

Now

$$\hat{q}(t_1)\hat{q}(t_2) = \frac{1}{\sin^2 \omega T} [\hat{q}_T^2 \sin \omega t_1 \sin \omega t_2$$
$$+ \hat{q}_T \hat{q}_0 \sin \omega t_1 \sin \omega (T - t_2)$$
$$+ \hat{q}_0 \hat{q}_T \sin \omega (T - t_1) \sin \omega t_2$$
$$+ \hat{q}_0^2 \sin \omega (T - t_2) \sin \omega (T - t_1)]$$

Thus using the solution of the second part of problem 10.1, it is easy to see that

$$\langle q_T | T(\hat{q}(t_1)\hat{q}(t_2)) | q_0 \rangle$$
$$= \left\{ \frac{\iota \hbar}{m\omega \sin \omega T} [\theta(t_1 - t_2) \sin \omega (T - t_1) \sin \omega t_2 \right.$$
$$+ \theta(t_2 - t_1) \sin \omega (T - t_2) \sin \omega t_1]$$
$$+ \frac{1}{\sin^2 \omega T} [q_0 \sin \omega (T - t_1) + q_T \sin \omega t_1]$$
$$\left. \times [q_0 \sin \omega (T - t_2) + q_T \sin \omega t_2] \right\} \langle q_F | q_I \rangle$$

Q10.8 For

$$H = \frac{\hat{p}^2}{2m} - \beta \hat{q}$$

with the boundary conditions

$$q(t) = q_f, \quad q(0) = q_i$$

show that

$$S_{cl} = \frac{\beta q_f}{2}t - \frac{\beta^2}{24m}t^3 + \frac{\beta q_i}{2}t + \frac{m(q_f - q_i)^2}{2t}.$$

Show that the propagator $K(q_f, t; q_i, 0)$ is

$$K(q_f, t; q_i, 0) = \left(\frac{m}{2\pi \hbar \iota t}\right)^{\frac{1}{2}} \exp\left(\frac{\iota S_{cl}}{\hbar}\right).$$

Solution: In Eqs. (10.152) of the text, put

$$J(t) = \beta$$

Then

$$I(t) = \beta(T - t), \quad I(0) = \beta T$$

$$\int_0^T dt\, I(t) = \frac{1}{2}\beta T^2, \quad \int_0^T dt\, I^2(t) = \frac{\beta^2}{3}T^3$$

Then

$$F = \frac{i}{\hbar}\left[\frac{m(q_F - q_I)^2}{2T} + \beta\frac{q_F - q_I}{2T}\beta T^2\right.$$

$$\left. + \frac{1}{2mT}\frac{1}{4}\beta^2 T^4 + q_I \beta T - \frac{1}{2m}\frac{\beta^2}{3}T^3\right]$$

$$= \frac{i}{\hbar}\left[\beta\frac{q_f}{2}T - \frac{\beta^2}{24m}T^3 + \beta\frac{q_I}{2}T + \frac{m(q_F - q_I)^2}{2T}\right]$$

$$\text{(10.12)}$$

with $T = t, q_F = q_f, q_I = q_i$ for the present problem. Thus from (10.152a)

$$K(q_f, t; q_i, 0) = \left(\frac{m}{2\pi \hbar \iota t}\right)^{1/2} \exp[iS_{cl}/\hbar]$$

where S_{cl} stands for [ ] in Eq. (10.12). We now calculate S_{cl}. Classically the given Hamiltonian gives

$$L = \frac{1}{2}m\dot{q}^2 + \beta q$$

and the equation of motion

$$m\ddot{q} = \beta$$

which has the solution

$$q(t) = \frac{\beta}{2m}t^2 + \left(\frac{q_f - q_i}{t_f} - \frac{\beta}{m}\frac{t_f}{2}\right)t + q_i$$

$$\dot{q}(t) = \frac{\beta t}{m} + \left(\frac{q_f - q_i}{t_f} - \frac{\beta}{m}\frac{t_f}{2}\right)$$

Thus

$$S_{cl} = \int_0^{t_f} L(q,\dot{q})dt = \frac{m}{2}\int_0^{t_f} \dot{q}^2 dt + \int_0^{t_f} qdt$$

$$= \frac{\beta^2}{24m}t_f^3 + \frac{m(q_f - q_i)^2}{2t_f} - \frac{\beta^2}{12m}t_f^3$$

$$+ \beta\frac{q_f - q_i}{2}t_f + q_i t_f$$

which is the identical with expression in the [ ] in Eq. (10.12) for $T = t_f = t$.

Chapter 11

Angular Momentum and Spin

Q11.1 Consider an operator $\hat{u}$, which obeys the commutation relations

$$[\hat{u}, J_z] = \frac{1}{2}\hbar\hat{u}$$

$$[\hat{u}, J_+] = 0.$$

Show that

$$\hat{u}|j, j\rangle = \text{constant} \quad |j - 1/2, j - 1/2\rangle.$$

Solution:

$$[\hat{u}, J_z] = \frac{1}{2}\hbar\hat{u}$$

$$[\hat{u}, J_+] = 0$$

Thus

$$J_z\hat{u}|j, j\rangle = \left(\hat{u}J_z - \frac{1}{2}\hbar\hat{u}\right)|j, j\rangle$$

$$= \left(j - \frac{1}{2}\right)\hbar\hat{u}|j, j\rangle$$

and

$$J_z^2\hat{u}|j, j\rangle = (j - 1/2)\hbar J_z\hat{u}|j, j\rangle$$

$$= (j - 1/2)^2\hbar^2\hat{u}|j, j\rangle \tag{11.1}$$

Now

$$[J_+, J_-] = 2\hbar J_z$$

$$J^2 = \frac{1}{2}(J_+ J_- + J_- J_+) + J_z^2$$

$$= J_- J_+ + \hbar J_z + J_z^2 \qquad (11.2)$$

Hence from Eqs. (11.1), (11.1), (11.2) and $J_+|j, j\rangle = 0$, we get

$$J^2 \hat{u}|j, j\rangle = \hbar^2 (j - 1/2)(j - 1/2 + 1)\hat{u}|j, j\rangle$$

Thus if $\hat{u}|j, j\rangle \neq 0$, it is an eigenstate of J^2 with eigenvalue $(j - \frac{1}{2})(j + \frac{1}{2})\hbar^2$. Hence

$$\hat{u}|j, j\rangle = \text{constant} \left|j - \frac{1}{2}, j - \frac{1}{2}\right\rangle$$

Q11.2 Show that
(i) $\sigma_x \sigma_y \sigma_z = i$.
(ii) $(\boldsymbol{\sigma} \cdot \mathbf{a})(\boldsymbol{\sigma} \cdot \mathbf{b}) = (\mathbf{a} \cdot \mathbf{b})I + i\boldsymbol{\sigma} \cdot (\mathbf{a} \times \mathbf{b})$, where $\mathbf{a}$ and $\mathbf{b}$ are ordinary vectors and I is a 2×2 unit matrix.
(iii) $(\boldsymbol{\sigma} \cdot \mathbf{n})$ has eigenvalues ± 1, where $\mathbf{n}$ is a unit vector.

Solution: (i) Now

$$\sigma_y \sigma_z - \sigma_z \sigma_y = 2i\sigma_x$$

$$\sigma_y \sigma_z + \sigma_z \sigma_y = 0$$

$$\sigma_y \sigma_z = i\sigma_x$$

Hence

$$\sigma_x \sigma_y \sigma_z = i\sigma_x^2 = i$$

(ii)

$$(\boldsymbol{\sigma} \cdot \mathbf{a})(\boldsymbol{\sigma} \cdot \mathbf{b}) = \sigma_i \sigma_j a_i b_j$$

$$\sigma_i \sigma_j = \delta_{ij} + i\epsilon_{ijk}\sigma_k$$

$$(\boldsymbol{\sigma} \cdot \mathbf{a})(\boldsymbol{\sigma} \cdot \mathbf{b}) = (\delta_{ij} + i\epsilon_{ijk}\sigma_k)a_i b_j$$

$$= \boldsymbol{a} \cdot \boldsymbol{b} + i\sigma_k (\boldsymbol{a} \times \boldsymbol{b})_k$$
$$= \boldsymbol{a} \cdot \boldsymbol{b} + i\boldsymbol{\sigma} \cdot (\boldsymbol{a} \times \boldsymbol{b})$$

(iii) From (ii) for $\boldsymbol{a} = \boldsymbol{b} = \boldsymbol{n}$

$$(\boldsymbol{\sigma} \cdot \boldsymbol{n})^2 = (\boldsymbol{\sigma} \cdot \boldsymbol{n})(\boldsymbol{\sigma} \cdot \boldsymbol{n}) = \boldsymbol{n} \cdot \boldsymbol{n} + i\boldsymbol{\sigma} \cdot (\boldsymbol{n} \times \boldsymbol{n})$$
$$= n^2 = 1$$

Thus $(\boldsymbol{\sigma} \cdot \boldsymbol{n})$ has eigenvalues ± 1

Q11.3 The eigenvectors of σ_z corresponding to eigenvalues ± 1 are denoted by $|\pm 1/2\rangle$. Show that

(i) $\frac{1}{\sqrt{2}}(|+1/2\rangle \pm |-1/2\rangle)$ are normalized eigenvectors of σ_x with eigenvalues ± 1.

(ii) $\frac{1}{\sqrt{2}}(|+1/2\rangle \pm i|-1/2\rangle)$ are normalized eigenvectors of σ_y with eigenvalues ± 1. Write the above eigenvectors as column vectors.

Solution: (i)

$$\sigma_+ = \sigma_x + i\sigma_y$$
$$\sigma_- = \sigma_x - i\sigma_y$$
$$\sigma_x = \frac{1}{2}(\sigma_+ + \sigma_-), \quad \sigma_y = -\frac{i}{2}(\sigma_+ - \sigma_-)$$

Thus (cf Problem 11.4, $\vec{S} = \frac{\hbar}{2}\vec{\sigma}$)

$$\sigma_x \left[\frac{1}{\sqrt{2}} \left(\left| \frac{1}{2} \right\rangle \pm \left| -\frac{1}{2} \right\rangle \right) \right]$$

$$= \frac{1}{2\sqrt{2}}(\sigma_+ + \sigma_-) \left[\left(\left| \frac{1}{2} \right\rangle \pm \left| -\frac{1}{2} \right\rangle \right) \right]$$

$$= \frac{2}{2\sqrt{2}} \left[\pm \left| \frac{1}{2} \right\rangle + \left| -\frac{1}{2} \right\rangle \right]$$

$$= \pm \frac{1}{\sqrt{2}} \left[\left(\left| \frac{1}{2} \right\rangle \pm \left| -\frac{1}{2} \right\rangle \right) \right]$$

$$\sigma_y \left[\frac{1}{\sqrt{2}} \left(\left| \frac{1}{2} \right\rangle \pm i \left| -\frac{1}{2} \right\rangle \right) \right]$$

$$= -\frac{i}{2\sqrt{2}}(\sigma_+ - \sigma_-) \left[\left(\left| \frac{1}{2} \right\rangle \pm i \left| -\frac{1}{2} \right\rangle \right) \right]$$

$$= -\frac{i}{\sqrt{2}} \left[\pm i \left(\left| \frac{1}{2} \right\rangle - \left| -\frac{1}{2} \right\rangle \right) \right]$$

$$= \frac{1}{\sqrt{2}} \left[\pm \left| \frac{1}{2} \right\rangle + i \left| -\frac{1}{2} \right\rangle \right]$$

$$= \pm \frac{1}{\sqrt{2}} \left[\left| \frac{1}{2} \right\rangle \pm i \left| -\frac{1}{2} \right\rangle \right]$$

(ii)

$$\frac{1}{\sqrt{2}} \left(\left| \frac{1}{2} \right\rangle \pm \left| -\frac{1}{2} \right\rangle \right) = \frac{1}{\sqrt{2}} \begin{pmatrix} 1 \\ \pm 1 \end{pmatrix}$$

$$\frac{1}{\sqrt{2}} \left(\left| \frac{1}{2} \right\rangle \pm i \left| -\frac{1}{2} \right\rangle \right) = \frac{1}{\sqrt{2}} \begin{pmatrix} 1 \\ \pm i \end{pmatrix}$$

Q11.4 S_x, S_y and S_z are components of spin $1/2$ operator $\mathbf{S}$ and obey the commutation relations

$$[S_x, S_y] = i\hbar S_z, \text{etc.}$$

Show that for $S_\pm = S_x \pm iS_y$

$$S_+| + 1/2 \rangle = 0$$
$$S_-| - 1/2 \rangle = 0$$
$$S_-| + 1/2 \rangle = \hbar| - 1/2 \rangle$$
$$S_+| - 1/2 \rangle = \hbar| + 1/2 \rangle,$$

where $|\pm 1/2 \rangle$ are eigenstates of S_z belonging to eigenvalues $\pm 1/2 \, \hbar$.

Solution:

$$[S_+, S_-] = 2\hbar S_z$$
$$[S_z, S_\pm] = \pm \hbar S_\pm$$

Thus

$$S_z S_\pm |1/2\rangle = (\pm\hbar S_\pm + S_\pm S_z)|1/2\rangle$$

$$= \left(\pm\hbar S_\pm + \frac{1}{2}\hbar S_\pm\right)|1/2\rangle$$

Hence $S_+| + \frac{1}{2}\rangle$ must be zero, since otherwise $S_+| + \frac{1}{2}\rangle$ will be an eigenstate of S_z with eigenvalue $\frac{3\hbar}{2}$ but S_z has only two eigenvalues $\pm\frac{\hbar}{2}$. On the other hand, unless $S_-|1/2\rangle = 0$, it has an eigenvalue $-\frac{1}{2}\hbar$, which is allowed. Thus

$$S_+|1/2\rangle = 0$$

$$S_-|1/2\rangle = \text{constant} \, |-1/2\rangle = C_- \left|-\frac{1}{2}\right\rangle$$

Similarly, we have

$$S_- \left|-\frac{1}{2}\right\rangle = 0$$

$$S_+ \left|-\frac{1}{2}\right\rangle = C_+ \left|\frac{1}{2}\right\rangle$$

Now

$$\left\langle -\frac{1}{2}\right| S_+^\dagger = \left\langle \frac{1}{2}\right| C_+^*$$

or

$$\left\langle -\frac{1}{2}\right| S_- = \left\langle \frac{1}{2}\right| C_+^*$$

Thus

$$\left\langle -\frac{1}{2}\right| S_- S_+ \left|-\frac{1}{2}\right\rangle = |C_+|^2 \left\langle \frac{1}{2}\bigg|\frac{1}{2}\right\rangle = |C_+|^2$$

Now

$$S^2 = \frac{1}{2}(S_+ S_- + S_- S_+) + S_z^2$$

$$= S_- S_+ + \hbar S_z + S_z^2$$

$$\therefore \quad S_- S_+ = S^2 - \hbar S_z - S_z^2$$

Thus

$$\left\langle -\frac{1}{2} \middle| S_- S_+ \middle| -\frac{1}{2} \right\rangle = \frac{3}{4}\hbar^2 + \frac{1}{2}\hbar^2 - \frac{1}{4}\hbar^2$$

$$= \hbar^2 \qquad (11.3)$$

Hence

$$|C_+|^2 = \hbar^2$$

and $C_+ = \hbar$, if we select C_+ to be real. Similarly $C_- = \hbar$. Hence

$$S_+ \left| -\frac{1}{2} \right\rangle = \hbar \left| \frac{1}{2} \right\rangle$$

$$S_- \left| \frac{1}{2} \right\rangle = \hbar \left| -\frac{1}{2} \right\rangle$$

Q11.5 If a particle is in an eigenstate of σ_x, find the probability of finding it in the eigenstate of σ_z belonging to eigenvalue $+1$.

Solution: Let $|\chi_\pm\rangle$ be eigenstates of σ_x with eigenvalues ± 1 respectively. They form a complete set so that an arbitrary state can be expanded in terms of them:

$$|\chi\rangle = a|\chi_+\rangle + b|\chi_-\rangle$$

Thus

$$a = \langle \chi_+ | \chi \rangle, \quad b = \langle \chi_- | \chi \rangle$$

If the particle is in an eigenstate of σ_x with eigenvalue $+1$, the probability of finding it in state $|\chi\rangle$ is

$$|a|^2 = |\langle \chi_+ | \chi \rangle|^2$$

In particular for the particle to be in an eigenstate of σ_z with eigenvalue 1, we have

$$a = \left\langle \chi_+ \middle| \frac{1}{2} \right\rangle = \frac{1}{\sqrt{2}} \left[\left\langle \frac{1}{2} \middle| \frac{1}{2} \right\rangle + \left\langle -\frac{1}{2} \middle| \frac{1}{2} \right\rangle \right] = \frac{1}{\sqrt{2}}$$

Hence the probability for the particle to be in an eigenstate of σ_z with eigenvalue 1, when the the particle is an eigenstate of σ_x with eigenvalue $+1$, is $\frac{1}{2}$.

Q11.6 Consider an electron in a uniform magnetic field **B** in the positive z-direction. The result of a measurement has shown that the electron spin is along the positive x-direction at $t = 0$. An arbitrary spin state at time t can be written as

$$|\chi(t)\rangle = a(t)|+1/2\rangle + b(t)|-1/2\rangle,$$

where $|\chi(t)\rangle$ satisfies the Schrödinger equation

$$i\hbar\frac{d}{dt}|\chi(t)\rangle = H|\chi(t)\rangle.$$

Find the probability at $t > 0$, for finding the electron in the spin state (a) $S_x = (1/2)\hbar$, (b)$S_x = -(1/2)\hbar$, and (c)$S_z = (1/2)\hbar$. Hint: As far as spin is concerned, the Hamiltonian $H = \mu_0 B\sigma_z$, where μ_0 is the magnetic moment of the electron. Use matrix representation for S_z etc.

Solution:

$$H = -\boldsymbol{\mu} \cdot \mathbf{B}$$

For electrons

$$\boldsymbol{\mu} = -\frac{e\hbar}{2mc}\boldsymbol{\sigma}$$

$$H = \frac{e\hbar}{2mc}\boldsymbol{\sigma} \cdot \mathbf{B}$$

$$= \frac{e\hbar}{2mc}B\sigma_z$$

Now

$$i\hbar\frac{d}{dt}|\chi(t)\rangle = \frac{e\hbar}{2mc}B\sigma_z|\chi(t)\rangle$$

$$\frac{d}{dt}|\chi(t)\rangle = -i\frac{eB}{2mc}\sigma_z|\chi(t)\rangle$$

$$= -i\omega\sigma_z \left[a(t) \left| \frac{1}{2} \right\rangle + b(t) \left| -\frac{1}{2} \right\rangle \right], \quad \omega = \frac{eB}{2m_c}$$

$$= -i\omega \left[a(t) \left| \frac{1}{2} \right\rangle - b(t) \left| -\frac{1}{2} \right\rangle \right] \quad\quad (11.4)$$

Now

$$a(t) = \left\langle \frac{1}{2} \middle| \chi(t) \right\rangle, \quad b(t) = \left\langle -\frac{1}{2} \middle| \chi(t) \right\rangle$$

Hence from Eq. (11.4), we have

$$\frac{da(t)}{dt} = -i\omega a(t)$$

$$\frac{db(t)}{dt} = i\omega b(t)$$

giving

$$a(t) = a(0)e^{-i\omega t}$$

$$b(t) = b(0)e^{i\omega t}$$

Hence

$$|\chi(t)\rangle = a(0)e^{-i\omega t} \left| +\frac{1}{2} \right\rangle + b(0)e^{i\omega t} \left| -\frac{1}{2} \right\rangle$$

$$|\chi(0)\rangle = a(0) \left| +\frac{1}{2} \right\rangle + b(0) \left| -\frac{1}{2} \right\rangle$$

At $t = 0$ the electron is in the state

$$|\chi_+\rangle = \frac{1}{\sqrt{2}} \left[\left| +\frac{1}{2} \right\rangle + \left| -\frac{1}{2} \right\rangle \right]$$

Thus

$$a(0) = b(0) = \frac{1}{\sqrt{2}}$$

and

$$|\chi(t)\rangle = \frac{1}{\sqrt{2}} \left[e^{-i\omega t} \left|+\frac{1}{2}\right\rangle + e^{i\omega t} \left|-\frac{1}{2}\right\rangle \right]$$

(a) Hence the probability of finding the electron in spin state $S_x = \frac{1}{2}\hbar$ is

$$|\langle\chi_+|\chi(t)\rangle|^2 = \frac{1}{4}|e^{-i\omega t} + e^{i\omega t}|^2$$

$$= \frac{1}{4}4\cos^2\omega t = \cos^2\omega t$$

(b) The probability to find it in the spin state with $S_x = -\frac{1}{2}\hbar$ at t :

$$|\langle\chi_-|\chi(t)\rangle|^2 = \frac{1}{4}|e^{-i\omega t} - e^{i\omega t}|^2 = \sin^2\omega t$$

(c) The probability to find it in the state with $S_z = \frac{1}{2}\hbar$:

$$\left|\left\langle\frac{1}{2}\bigg|\chi(t)\right\rangle\right|^2 = \left|\frac{1}{\sqrt{2}}e^{-i\omega t}\right|^2 = \frac{1}{2}$$

Q11.7 Consider an operator $\hat{u}$ which obeys the commutation relations

$$[\hat{u}, J_z] = (1/2)\hbar\hat{u}$$

$$[[\hat{u}, J^2], J^2] = \frac{1}{2}(\hat{u}J^2 + J^2\hat{u}) + \frac{3\hbar^4}{16}\hat{u}.$$

Consider the representation in which the basis vectors are $|jm\rangle$, the simultaneous eigenvectors of J^2 and J_z belonging to the eigenvalues $j(j+1)\hbar^2$ and $m\hbar$ respectively. Show that if

$$\langle jm|\hat{u}|j'm'\rangle \neq 0$$

(i) $m' = m + 1/2$
(ii) $(j'(j'+1) - j(j+1))^2 = \frac{1}{2}(j'(j'+1) + j(j+1)) + \frac{3}{16}$.

Solution: (i)

$$\langle jm|\hat{u}|j'm'\rangle = \frac{2}{\hbar}\langle jm|[\hat{u}, J_z]|j'm'\rangle$$

$$= \frac{2}{\hbar}(m' - m)\hbar\langle jm|\hbar u|j'm'\rangle$$

$$\therefore \left[(m' - m) - \frac{1}{2}\right]\langle jm|u|j'm'\rangle = 0$$

Hence

$$m' = m + 1/2$$

(ii) R.H.S

$$\left\langle jm\left|\left[\frac{\hbar^2}{2}(\hat{u}J^2 + J^2\hat{u})\rangle + \frac{3\hbar^4}{16}\right]\right|j'm'\right\rangle$$

$$= \frac{1}{2}\left[j'(j' + 1)\hbar^4 + j(j + 1)\hbar^4 + \frac{3\hbar^4}{16}\right]\langle jm|\hat{u}|j'm'\rangle$$

L.H.S

$$\langle jm|\{[\hat{u}, J^2]J^2 - J^2[\hat{u}, J^2]\}|j'm'\rangle$$

$$= [j'(j' + 1) - j(j + 1)]\hbar^2\langle jm|[\hat{u}, J^2]|j'm'\rangle$$

$$= [j'(j' + 1) - j(j + 1)]\hbar^2[j'(j' + 1)$$

$$- j(j + 1)]\hbar^2\langle jm|\hat{u}|j'm'\rangle$$

$$= [j'(j' + 1) - j(j + 1)]^2\hbar^4\langle jm|\hat{u}|j'm'\rangle$$

Hence

$$[j'(j' + 1) - j(j + 1)]^2 = \frac{1}{2}[j'(j' + 1) - j(j + 1)] + \frac{3}{16}$$

Q11.8 Consider a system of two distinguishable particles, each with spin 1/2 and magnetic moment $\boldsymbol{\mu}_1 = \mu_1\boldsymbol{\sigma}_1, \boldsymbol{\mu}_2 = \mu_2\boldsymbol{\sigma}_2$ in an external magnetic field B in the z-direction. The spin-spin interaction of the particles is $b\boldsymbol{\sigma}_1 \cdot \boldsymbol{\sigma}_2$, where b is a constant. Find the exact energy eigenvalues of this system.

Solution:

$$H = -(\mu_1\boldsymbol{\sigma}_1 + \mu_2\boldsymbol{\sigma}_2) \cdot \mathbf{B} + b\boldsymbol{\sigma}_1 \cdot \boldsymbol{\sigma}_2$$

$$= -\frac{1}{2}\left[(\mu_1 + \mu_2)(\boldsymbol{\sigma}_1 + \boldsymbol{\sigma}_2) + (\mu_1 - \mu_2)(\boldsymbol{\sigma}_1 - \boldsymbol{\sigma}_2)\right] \cdot \mathbf{B}$$

$$+b\boldsymbol{\sigma}_1 \cdot \boldsymbol{\sigma}_2$$

$\mathbf{B} = B\mathbf{e}_z$, then

$$H = -\frac{B}{2}[(\mu_1 + \mu_2)(\sigma_{1z} + \sigma_{2z})$$

$$+(\mu_1 - \mu_2)(\sigma_{1z} - \sigma_{2z})] + b\boldsymbol{\sigma}_1 \cdot \boldsymbol{\sigma}_2$$

The spin wave functions for two particle system are

$$S = 0 : \chi_-^0 = \frac{1}{\sqrt{2}}[\chi_1(\uparrow)\chi_2(\downarrow) - \chi_1(\downarrow)\chi_2(\uparrow)]$$

$$\times \text{ spin singlet state}$$

Spin triplet states:

$$S = 1 : \quad \chi_+^{+1} = \chi_1(\uparrow)\chi_2(\uparrow)$$

$$\chi_+^0 = \frac{1}{\sqrt{2}}[\chi_1(\uparrow)\chi_2(\downarrow) + \chi_1(\downarrow)\chi_2(\uparrow)]$$

$$\chi_+^{-1} = \chi_1(\downarrow)\chi_2(\downarrow)$$

Now

$$\boldsymbol{\sigma}_1 \cdot \boldsymbol{\sigma}_2 = -3 \quad \text{for spin singlet state}$$
$$= 1 \quad \text{for spin triplet state}$$

Hence

$$H\chi_-^0 = -B(\mu_1 - \mu_2)\chi_+^0 - 3b\chi_-^0$$
$$H\chi_+^0 = -B(\mu_1 - \mu_2)\chi_-^0 + b\chi_+^0$$
$$H\chi_+^{+1} = [-B(\mu_1 + \mu_2) + b]\chi_+^{+1}$$
$$H\chi_+^{-1} = [B(\mu_1 + \mu_2) + b]\chi_+^{-1}$$

Hence the eigenvalues for the spin triplet states χ_+^{+1}, χ_+^{-1} are

$$E_{\pm 1} = \mp(\mu_1 + \mu_2)B + b$$

For $S_z = 0$, the Hamiltionian is a 2×2 matrix i.e.

$$H = \begin{pmatrix} -3b & -B(\mu_1 - \mu_2) \\ -B(\mu_1 - \mu_2) & b \end{pmatrix}$$

Hence the eigenvalues are given by

$$\begin{vmatrix} -3b - E & -B(\mu_1 - \mu_2) \\ -B(\mu_1 - \mu_2) & b - E \end{vmatrix} = 0$$

or

$$E^2 + 2bE - B^2(\mu_1 - \mu_2)^2 - 3b^2 = 0$$

or

$$E_{\pm} = -b \pm \sqrt{4b^2 + B^2(\mu_1 - \mu_2)^2}$$

Q11.9 In the normal Zeeman effect, an energy level characterized by quantum number n is split into $(2l+1)$ different levels:

$$E_{n,m} = E_n + \frac{e\hbar}{2m_e c}Bm,$$

where $m = -l, \ldots, l$. The equal spacing between the levels is given by

$$\Delta E = \frac{e\hbar}{2m_e c}B.$$

Show that

$$\Delta E = \left(\frac{1}{2}\alpha^2 m_e c^2\right)\alpha\frac{B}{e/a^2}$$

$$= \frac{B}{(2.4 \times 10^9 \text{Gauss})} \times 13.6\text{eV},$$

where $\alpha = \frac{e^2}{\hbar c} = \frac{1}{137}$ is the fine structure constant and $E_1 = \frac{1}{2}m_e c^2 \alpha^2 = 13.6$ eV is the binding energy of electron in the first Bohr orbit of hydrogen atom. a is the radius of first Bohr orbit. Draw the energy level diagram for the Zeeman effect for p and d level. In atomic transitions $\Delta m = m_f - m_i = -1, 0, 1$. Draw 8 possible transitions on the energy level diagram.

Solution: Now

$$\Delta E = \frac{e\hbar}{2m_e c^2} B$$

Bohr radius : $a^2 = \frac{\hbar^2 c^2}{(m_e c^2)^2 \alpha^2} \approx (0.5 \times 10^{-8} \text{cm})^2$

$$(11.5)$$

Writing ΔE in the form

$$\Delta E = \frac{\hbar}{2m_e c} \frac{e^2}{a^2} \frac{B}{(e/a^2)}$$

and using

$$\frac{e^2}{a^2} = (c\hbar\alpha) \frac{(m_e c^2)^2 \alpha^2}{c^2 \hbar^2}$$

we get

$$\Delta E = \left(\frac{1}{2} \alpha^2 m_e c^2 \right) \frac{\alpha B}{(e/a^2)}$$

Now

$$\frac{1}{2} \alpha^2 m_e c^2) \approx 13.6 \text{eV}$$

Bohr radius $a \approx 0.53 \times 10^{-8}$cm, $e = 4.8 \times 10^{-10}$esu

$$\frac{e}{\alpha a^2} = 2.34 \times 10^9 \text{Gauss}$$

Hence

$$\Delta E = \frac{B}{2.34 \times 10^9 \text{Gauss}} \times 13.6 \text{eV}$$

Now

$$E_{n,m} = E_n - m\Delta E$$

$$\text{p-state} : \quad \ell = 1, m = 1, 0, -1$$

$$\text{d-state} : \quad \ell = 2, m = 2, 1, 0, -1, -2$$

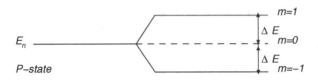

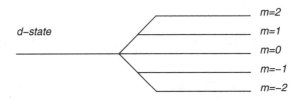

Q11.10 An electron can be regarded as a magnetic dipole of magnetic moment $\boldsymbol{\mu} = \mu_0 \boldsymbol{\sigma}$, where $\mu_0 = \frac{e\hbar}{2m_e c}$ is the electron magnetic moment. The interaction energy between magnetic dipoles is given by

$$V = \frac{1}{r^3} \left\{ (\boldsymbol{\mu}_1 \cdot \boldsymbol{\mu}_2) - 3 \frac{(\boldsymbol{\mu}_1 \cdot \mathbf{r})(\boldsymbol{\mu}_2 \cdot \mathbf{r})}{r^2} \right\}.$$

Find the dipole-dipole magnetic interaction energy of an electron and a positron at a fixed distance a, in eigenstates of total spin.

Solution:

$$V = \frac{1}{r^3} \left[\boldsymbol{\mu}_1 \cdot \boldsymbol{\mu}_2 - 3 \frac{(\boldsymbol{\mu}_1 \cdot \mathbf{r})(\boldsymbol{\mu}_2 \cdot \mathbf{r})}{r^2} \right]$$

$$\boldsymbol{\mu}_1 = \mu_0 \boldsymbol{\sigma}_1, \ \boldsymbol{\mu}_2 = \mu_0 \boldsymbol{\sigma}_2$$

$$V = \frac{\mu_0}{r^3} \left[\boldsymbol{\sigma}_1 \cdot \boldsymbol{\sigma}_2 - 3 \frac{(\boldsymbol{\sigma}_1 \cdot \mathbf{r})(\boldsymbol{\sigma}_2 \cdot \mathbf{r})}{r^3} \right]$$

Eigenstates for the total spin $s = 0$ and $s = 1$ are

$$s = 0 \quad \chi_-^0 = \frac{1}{\sqrt{2}}(\chi_+(1)\chi_-(2) - \chi_-(1)\chi_+(2))$$

$$s = 1 \quad \chi_+^{+1} = \chi_+(1)\chi_+(2)$$

$$\chi_+^0 = \frac{1}{\sqrt{2}}(\chi_+(1)\chi_-(2) + \chi_-(1)\chi_+(2))$$

$$\chi_+^{-1} = \chi_-(1)\chi_-(2)$$

For simplicity take $\mathbf{r}$ along z-axis $\mathbf{r} = re_z$

$$\boldsymbol{\sigma}_1 \cdot \mathbf{r} = r\sigma_{1z}$$

$$\boldsymbol{\sigma}_2 \cdot \mathbf{r} = r\sigma_{2z}$$

Thus

$$V = \frac{\mu_0}{r^3}[\boldsymbol{\sigma}_1 \cdot \boldsymbol{\sigma}_2 - 3\sigma_{1z}\sigma_{2z}]$$

$$\mathbf{S}^2 = \frac{\hbar^2}{4}(\boldsymbol{\sigma}_1 + \boldsymbol{\sigma}_2)^2$$

$$= \frac{\hbar^2}{4}[\sigma_1^2 + \sigma_2^2 + 2\boldsymbol{\sigma}_1 \cdot \boldsymbol{\sigma}_2]$$

$$= \hbar^2\left[\frac{3}{2} + \frac{1}{2}\boldsymbol{\sigma}_1 \cdot \boldsymbol{\sigma}_2\right]$$

so that

$$\boldsymbol{\sigma}_1 \cdot \boldsymbol{\sigma}_2 = \left(\frac{2S^2}{\hbar^2} - 3\right)$$

Eigenvalues of S^2 are $s(s+1)\hbar^2$ where $s = 0$ or $s = 1$

$$\boldsymbol{\sigma}_1 \cdot \boldsymbol{\sigma}_2\chi_-^0 = -3\chi_-^0$$

$$\sigma_{1z}\sigma_{2z}\chi_-^0(1,2) = -\chi_-^0$$

$$\boldsymbol{\sigma}_1 \cdot \boldsymbol{\sigma}_2\chi_+^{+1} = \chi_+^{+1}$$

$$\boldsymbol{\sigma}_1 \cdot \boldsymbol{\sigma}_2\chi_+^0 = +\chi_+^0$$

$$\boldsymbol{\sigma}_1 \cdot \boldsymbol{\sigma}_2\chi_+^{-1} = +\chi_+^{-1}$$

$$\sigma_{1z}\sigma_{2z}\chi_+^{+1} = \chi_+^{+1}$$

$$\sigma_{1z}\sigma_{2z}\chi_+^0 = -\chi_+^0$$

$$\sigma_{1z}\sigma_{2z}\chi_+^{-1} = \chi_+^{-1}$$

Therefore magnetic interaction energy

$$\langle \chi_-^0 | V | \chi_-^0 \rangle = \left\langle \chi_-^0 \left| \frac{\mu_0}{r^3} (\boldsymbol{\sigma}_1 \cdot \boldsymbol{\sigma}_2 - 3\sigma_{1z}\sigma_{2z}) \right| \chi_-^0 \right\rangle$$

$$= \left\langle \chi_-^0 \left| \frac{\mu_0}{r^3} (-3 + 3) \right| \chi_-^0 \right\rangle = 0$$

$$\langle \chi_+^{+1} | V | \chi_+^{+1} \rangle = -\frac{2\mu_0}{r^3}$$

$$\langle \chi_+^0 | V | \chi_+^0 \rangle = \frac{4\mu_0}{r^3}$$

$$\langle \chi_+^{-1} | V | \chi_+^{-1} \rangle = -\frac{2\mu_0}{r^3}$$

Q11.11 Show explicitly that the state $|j = l - 1/2, m\rangle$ as given by Eq. (11.92) is an eigenstate of the operator $(\mathbf{L} + \mathbf{S})^2$ with eigenvalue $j(j + 1)\hbar^2$.

Solution: From Eq. (11.92) of the text

$$J^2 \left| \ell - \frac{1}{2}, m \right\rangle = -\sqrt{\frac{\ell - m + \frac{1}{2}}{2\ell + 1}} (\mathbf{L}^2 + \mathbf{S}^2 + 2\mathbf{L} \cdot \mathbf{S})$$

$$\times \left| \ell, m - \frac{1}{2} \right\rangle \left| \frac{1}{2}, \frac{1}{2} \right\rangle$$

$$+ \sqrt{\frac{\ell + m + \frac{1}{2}}{2\ell + 1}} (\mathbf{L}^2 + \mathbf{S}^2 + 2\mathbf{L} \cdot \mathbf{S})$$

$$\times \left| \ell, m + \frac{1}{2} \right\rangle \left| \frac{1}{2}, -\frac{1}{2} \right\rangle$$

Now

$$L_+ \left| \ell, m - \frac{1}{2} \right\rangle = \hbar \sqrt{\left(\ell + m + \frac{1}{2} \right) \left(\ell - m + \frac{1}{2} \right)} \left| \ell, m + \frac{1}{2} \right\rangle$$

$$L_- \left| \ell, m + \frac{1}{2} \right\rangle = \hbar \sqrt{\left(\ell + m + \frac{1}{2} \right) \left(\ell - m + \frac{1}{2} \right)} \left| \ell, m - \frac{1}{2} \right\rangle$$

$$S_+ \left| \frac{1}{2}, \frac{1}{2} \right\rangle = 0$$

$$S_+ \left| \frac{1}{2}, -\frac{1}{2} \right\rangle = \hbar \left| \frac{1}{2}, \frac{1}{2} \right\rangle$$

$$S_- \left| \frac{1}{2}, \frac{1}{2} \right\rangle = \hbar \left| \frac{1}{2}, -\frac{1}{2} \right\rangle$$

$$S_- \left| \frac{1}{2}, -\frac{1}{2} \right\rangle = 0$$

$$2\mathbf{L} \cdot \mathbf{S} = L_+ S_- + L_- S_+ + 2 L_z S_z$$

$$\therefore J^2 \left| \ell - \frac{1}{2}, m \right\rangle = -\sqrt{\frac{\ell - m + \frac{1}{2}}{2\ell + 1}}$$

$$\times \left[\left(\ell(\ell+1)\hbar^2 + \frac{3}{4} \right) \left| \ell, m - \frac{1}{2} \right\rangle \left| \frac{1}{2}, \frac{1}{2} \right\rangle \right.$$

$$+ \hbar^2 \sqrt{\left(\ell + m + \frac{1}{2} \right) \left(\ell - m + \frac{1}{2} \right)}$$

$$\times \left| \ell, m + \frac{1}{2} \right\rangle \left| \frac{1}{2}, -\frac{1}{2} \right\rangle$$

$$\left. + 2 \left(m - \frac{1}{2} \right) \frac{1}{2} \hbar^2 \left| \ell, m - \frac{1}{2} \right\rangle \left| \frac{1}{2}, \frac{1}{2} \right\rangle \right]$$

$$+ \sqrt{\frac{\ell + m + \frac{1}{2}}{2\ell + 1}} \left[\left(\ell(\ell+1)\hbar^2 + \frac{3}{4} \right) \left| \ell, m + \frac{1}{2} \right\rangle \left| \frac{1}{2}, -\frac{1}{2} \right\rangle \right.$$

$$+ \hbar^2 \sqrt{\left(\ell + m + \frac{1}{2} \right) \left(\ell - m + \frac{1}{2} \right)}$$

$$\times \left| \ell, m - \frac{1}{2} \right\rangle \left| \frac{1}{2}, \frac{1}{2} \right\rangle$$

$$\left. + 2 \left(m + \frac{1}{2} \right) \left(-\frac{1}{2} \right) \hbar^2 \left| \ell, m + \frac{1}{2} \right\rangle \left| \frac{1}{2}, -\frac{1}{2} \right\rangle \right]$$

$$= \hbar^2 \left[l(l+1) + \frac{3}{4} + \left(m - \frac{1}{2} \right) - \left(l + m + \frac{1}{2} \right) \right]$$

$$\times \left[-\sqrt{\frac{l - m + \frac{1}{2}}{2l + 1}} \left| l, m - \frac{1}{2} \right\rangle \left| \frac{1}{2}, \frac{1}{2} \right\rangle \right]$$

$$+ \hbar^2 \left[-\left(l - m + \frac{1}{2} \right) + l(l+1) + \frac{3}{4} - \left(m + \frac{1}{2} \right) \right]$$

$$\times \left[\sqrt{\frac{l + m + \frac{1}{2}}{2l + 1}} \left| l, m + \frac{1}{2} \right\rangle \left| \frac{1}{2}, -\frac{1}{2} \right\rangle \right]$$

$$= \left(l^2 - \frac{1}{4} \right) \hbar^2 \left| l - \frac{1}{2}, m \right\rangle$$

Q11.12 Consider an electron in an external magnetic field **B**. The energy of the spinning electron is given by

$$H_{\text{spin}} = -\boldsymbol{\mu} \cdot \mathbf{B} = \frac{eg}{2mc} \mathbf{S} \cdot \mathbf{B},$$

show that

$$\frac{d\mathbf{S}}{dt} = -\frac{eg}{2mc} \mathbf{S} \times \mathbf{B}.$$

Further show that the expectation value of the spin operator **S** at time t, when initially the particle is in the spin state with $S_x = \frac{1}{2}\hbar$ is given by

$$\langle \mathbf{S} \rangle = \frac{\hbar}{2} \cos \omega t \, \mathbf{e}_x + \frac{\hbar}{2} \sin \omega t \, \mathbf{e}_y,$$

where $\mathbf{e}_x$ and $\mathbf{e}_y$ are the unit vectors along the x- and y-axis respectively and $\omega = \frac{egB}{2mc}$.

Solution:

$$H_{\text{spin}} = -\boldsymbol{\mu} \cdot \mathbf{B} = \frac{eg}{2mc} \mathbf{S} \cdot \mathbf{B}$$

Heisenberg equation of motion:

$$i\hbar \frac{d\mathbf{S}}{dt} = [\mathbf{S}, H] = \frac{eg}{2mc} [\mathbf{S}, \mathbf{S} \cdot \mathbf{B}]$$

Thus

$$\frac{dS_i}{dt} = -\frac{i}{\hbar}\frac{eg}{2mc}B_j[S_i, S_j]$$

$$= \frac{eg}{2mc}\epsilon_{ijk}B_j S_k$$

$$= -\frac{eg}{2mc}(\mathbf{S} \times \mathbf{B})_i$$

Hence

$$\frac{d\mathbf{S}}{dt} = -\frac{eg}{2mc}\mathbf{S} \times \mathbf{B}$$

Take **B** along z-axis:

$$\mathbf{B} = B\mathbf{e}_z$$

Thus

$$\frac{dS_i}{dt} = -\frac{egB}{2mc}\epsilon_{ijz}S_j$$

giving

$$\frac{dS_x}{dt} = -\omega S_y, \quad \frac{dS_y}{dt} = \omega S_x, \quad \frac{dS_z}{dt} = 0$$

Hence

$$\frac{d^2 S_x}{dt^2} = -\omega\frac{dS_y}{dt} = -\omega^2 S_x \tag{11.6}$$

$$\frac{d^2 S_y}{dt^2} = -\omega\frac{dS_x}{dt} = -\omega^2 S_y \tag{11.7}$$

$$S_z(t) = a \text{ constant } = S_z(0)$$

The solutions of Eqs. (11.6) and (11.7) are

$$S_x(t) = a_x e^{i\omega t} + b_x e^{-i\omega t}$$

$$S_y(t) = a_y e^{i\omega t} + b_y e^{-i\omega t}$$

Now the boundary condition gives

$$\langle S_x(0)\rangle = \frac{\hbar}{2}, \quad \langle S_y(0)\rangle = 0,$$

$$\left\langle\frac{dS_x}{dt}\right\rangle_{t=0} = 0, \quad \left\langle\frac{dS_y}{dt}\right\rangle_{t=0} = \omega\frac{\hbar}{2}$$

Thus

$$a_x = \frac{\hbar}{4} = b_x, \qquad a_y = -b_y = -\frac{i\hbar}{4}$$

$$S_z(0) = 0$$

Hence we have

$$\langle \mathbf{S} \rangle = \frac{\hbar}{2} \cos \omega t \, \mathbf{e}_x + \frac{\hbar}{2} \sin \omega t \, \mathbf{e}_y$$

Q11.13 The Hamiltonian for a particle of mass m is given by

$$H = \frac{\hat{p}^2}{2m} + V(r) + U\hbar \mathbf{L} \cdot \boldsymbol{\sigma},$$

where U is some constant. Show that $\mathbf{L}$ is not a constant of motion, but $\mathbf{J} = \mathbf{L} + \frac{1}{2}\hbar\boldsymbol{\sigma}$ is a constant of motion.

Solution:

$$H = \frac{\hat{p}^2}{2m} + V(r) + U\hbar \mathbf{L} \cdot \boldsymbol{\sigma},$$

Now from Heisenberg's equation of motion,

$$i\hbar \frac{dL_i}{dt} = [L_i, H]$$

$$i\hbar \frac{d\sigma_i}{dt} = [\sigma_i, H]$$

Now $\frac{\hat{p}^2}{2m} + V(r)$ is rotationally invariant, therefore L_i commutes with this part of the Hamiltionian. Thus

$$i\hbar \frac{dL_i}{dt} = U\hbar[L_i, L_j\sigma_j]$$

$$i\hbar \frac{d\sigma_i}{dt} = U\hbar[\sigma_i, L_j\sigma_j]$$

Therefore

$$\frac{dL_i}{dt} = -iU[L_i, L_j]\sigma_j$$

$$= (-i)(i\hbar\epsilon_{ijk}L_k)\sigma_j = U\hbar(\boldsymbol{\sigma} \times \mathbf{L})_i \neq 0$$

$$\frac{d\sigma_i}{dt} = -iUL_j[\sigma_i, \sigma_j]$$

$$= 2U\epsilon_{ijk}L_j\sigma_k = -2U(\boldsymbol{\sigma} \times \mathbf{L})_i \neq 0$$

Hence

$$\frac{d\mathbf{J}}{dt} = \frac{d}{dt}\left(\mathbf{L} + \frac{1}{2}\hbar\boldsymbol{\sigma}\right) = 0$$

i.e. $\mathbf{J}$ is a constant of motion.

Q11.14 Show that for the spin $\frac{1}{2}\hbar$ case $(\mathbf{J} = \mathbf{S} = \frac{1}{2}\hbar\boldsymbol{\sigma})$, the rotation operator can be written as

$$U_R(\theta) = e^{-(i/2)\theta\mathbf{n}\cdot\boldsymbol{\sigma}},$$

$$= \cos\frac{\theta}{2} - i\sin\frac{\theta}{2}\mathbf{n}\cdot\boldsymbol{\sigma},$$

where θ is the angle of rotation about $\mathbf{n}(n^2 = 1)$. In general a spin state can be written as

$$|\chi\rangle = C_1\left|+\frac{1}{2}\right\rangle + C_2\left|-\frac{1}{2}\right\rangle$$

$$= \begin{pmatrix} C_1 \\ C_2 \end{pmatrix},$$

called a two component spinor. Its transformation law under the rotation is

$$|\chi'\rangle = e^{-(i/2)\theta\mathbf{n}\cdot\boldsymbol{\sigma}}|\chi\rangle.$$

Show that
(i) for $\theta = 2\pi$, $|\chi\rangle = -|\chi\rangle$
(ii) for rotation about x-axis, for which $\hat{\mathbf{n}} = \mathbf{e}_x$, $\boldsymbol{\sigma}\cdot\hat{\mathbf{n}} = \sigma_x$,

$$\begin{pmatrix} C_1' \\ C_2' \end{pmatrix} = \begin{pmatrix} \cos\frac{\theta}{2} & -i\sin\frac{\theta}{2} \\ -i\sin\frac{\theta}{2} & \cos\frac{\theta}{2} \end{pmatrix}\begin{pmatrix} C_1 \\ C_2 \end{pmatrix}.$$

Compare it with the transformation law for a vector under the above rotation. For the same rotation, show that

$$\boldsymbol{\sigma}' = e^{-i(\theta/2)\mathbf{n}\cdot\boldsymbol{\sigma}}\boldsymbol{\sigma}\, e^{i(\theta/2)\mathbf{n}\cdot\boldsymbol{\sigma}}$$

gives

$$\sigma'_x = \sigma_x$$
$$\sigma'_y = \cos\theta\sigma_y + \sin\theta\sigma_z$$
$$\sigma'_z = -\sin\theta\sigma_y + \cos\theta\sigma_z,$$

i.e. $\boldsymbol{\sigma}$ transforms like a vector under the rotation.

Solution:

$$U_R(\theta) = e^{-\frac{i}{2}\theta\mathbf{n}\cdot\boldsymbol{\sigma}}$$

$$= 1 + \left(-\frac{i}{2}\theta\mathbf{n}\cdot\boldsymbol{\sigma}\right) + \frac{1}{2}\left(-\frac{i}{2}\theta\mathbf{n}\cdot\boldsymbol{\sigma}\right)^2$$

$$+ \frac{1}{3}\left(-\frac{i}{2}\theta\mathbf{n}\cdot\boldsymbol{\sigma}\right)^3 + \cdots$$

$$= 1 - \frac{i}{2}\theta(\mathbf{n}\cdot\boldsymbol{\sigma}) - \frac{1}{2}\left(\frac{\theta}{2}\right)^2(\mathbf{n}\cdot\boldsymbol{\sigma})^2$$

$$- \frac{i}{3!}\left(\frac{\theta}{2}\right)^3(\mathbf{n}\cdot\boldsymbol{\sigma})^3 + \cdots$$

Now

$$(\mathbf{n}\cdot\boldsymbol{\sigma})^2 = 1, \quad (\mathbf{n}\cdot\boldsymbol{\sigma})^3 = \mathbf{n}\cdot\boldsymbol{\sigma}$$

Hence

$$U_R(\theta) = \left(1 - \frac{1}{2!}\left(\frac{\theta}{2}\right)^2 + \cdots\right)$$

$$- i(\mathbf{n}\cdot\boldsymbol{\sigma})\left(\frac{\theta}{2} + \frac{1}{3!}\left(\frac{\theta}{2}\right)^3 + \cdots\right)$$

$$= \cos\frac{\theta}{2} - i\sin\frac{\theta}{2}\mathbf{n}\cdot\boldsymbol{\sigma}$$

Rotated spin state

$$|\chi'\rangle = e^{-\frac{i}{2}\theta\mathbf{n}\cdot\boldsymbol{\sigma}}|\chi\rangle$$

$$= \left(\cos\frac{\theta}{2} - i\sin\frac{\theta}{2}\mathbf{n}\cdot\boldsymbol{\sigma}\right)|\chi\rangle$$

(i) $\theta = 2\pi$

$$|\chi'\rangle = \cos \pi |\chi\rangle = -|\chi\rangle$$

(ii) $\mathbf{n} = \mathbf{e}_x, \ \mathbf{n} \cdot \boldsymbol{\sigma} = \sigma_x$

$$|\chi'\rangle = \left(\cos \frac{\theta}{2} - i \sin \frac{\theta}{2} \sigma_x \right) |\chi\rangle$$

Now

$$\sigma_x = \begin{pmatrix} 0 & 1 \\ 1 & 0 \end{pmatrix}$$

$$\therefore |\chi'\rangle = \begin{pmatrix} \cos \dfrac{\theta}{2} & -i \sin \dfrac{\theta}{2} \\ -i \sin \dfrac{\theta}{2} & \cos \dfrac{\theta}{2} \end{pmatrix} |\chi\rangle$$

Hence

$$\begin{pmatrix} C_1' \\ C_2' \end{pmatrix} = \begin{pmatrix} \cos \dfrac{\theta}{2} & -i \sin \dfrac{\theta}{2} \\ -i \sin \dfrac{\theta}{2} & \cos \dfrac{\theta}{2} \end{pmatrix} \begin{pmatrix} C_1 \\ C_2 \end{pmatrix} \tag{11.8}$$

$$\boldsymbol{\sigma}' = e^{-\frac{i}{2}\theta \boldsymbol{\sigma} \cdot \mathbf{n}} \boldsymbol{\sigma} e^{\frac{i}{2}\theta \boldsymbol{\sigma} \cdot \mathbf{n}}$$

$$= \left(\cos \frac{\theta}{2} - i \sin \frac{\theta}{2} \boldsymbol{\sigma} \cdot \mathbf{n} \right) \boldsymbol{\sigma} \left(\cos \frac{\theta}{2} + i \sin \frac{\theta}{2} \boldsymbol{\sigma} \cdot \mathbf{n} \right)$$

$$= \cos^2 \frac{\theta}{2} \boldsymbol{\sigma} + i \sin \frac{\theta}{2} \cos \frac{\theta}{2} (\boldsymbol{\sigma}\boldsymbol{\sigma} \cdot \mathbf{n} - \boldsymbol{\sigma} \cdot \mathbf{n}\boldsymbol{\sigma})$$

$$+ \sin^2 \frac{\theta}{2} \boldsymbol{\sigma} \cdot \mathbf{n}\boldsymbol{\sigma} \cdot \mathbf{n}$$

Now

$$(\boldsymbol{\sigma}\boldsymbol{\sigma} \cdot \mathbf{n} - \boldsymbol{\sigma} \cdot \mathbf{n}\boldsymbol{\sigma})_i = (\sigma_i \sigma_j - \sigma_j \sigma_i) n_j$$

$$= 2i \epsilon_{ijk} \sigma_k n_j = 2i (\mathbf{n} \times \boldsymbol{\sigma})_i$$

$$\boldsymbol{\sigma} \cdot \mathbf{n}\boldsymbol{\sigma} \cdot \mathbf{n} = 2(\boldsymbol{\sigma} \cdot \mathbf{n})\mathbf{n} - \boldsymbol{\sigma}$$

Hence

$$\boldsymbol{\sigma}' = \cos^2\frac{\theta}{2}\boldsymbol{\sigma} + 2i\sin\frac{\theta}{2}\cos\frac{\theta}{2}i(\mathbf{n}\times\boldsymbol{\sigma})$$

$$+\sin^2\frac{\theta}{2}(2(\boldsymbol{\sigma}\cdot\mathbf{n})\mathbf{n} - \boldsymbol{\sigma})$$

$$= \cos\theta\boldsymbol{\sigma} - \sin\theta(\mathbf{n}\times\boldsymbol{\sigma}) + (1 - \cos\theta)(\boldsymbol{\sigma}\cdot\mathbf{n})\mathbf{n}$$

In particular for $\mathbf{n} = \mathbf{e}_x$

$$\boldsymbol{\sigma}' = \cos\theta\boldsymbol{\sigma} - \sin\theta(\sigma_y\mathbf{e}_z - \sigma_z\mathbf{e}_y) + (1 - \cos\theta)\sigma_x\mathbf{e}_x$$

i.e.

$$\sigma_x' = \sigma_x$$

$$\sigma_y' = \cos\theta\sigma_y + \sin\theta\sigma_z$$

$$\sigma_z' = -\sin\theta\sigma_y + \cos\theta\sigma_z$$

Q11.15 Show that

$$\boldsymbol{\sigma}\cdot\mathbf{p}e^{-(i/2)\theta\mathbf{n}\cdot\boldsymbol{\sigma}} = e^{-(i/2)\theta\mathbf{n}\cdot\boldsymbol{\sigma}}\boldsymbol{\sigma}\cdot\mathbf{e}_z = e^{(i/2)\theta\mathbf{n}\cdot\boldsymbol{\sigma}}\sigma_z.$$

Hence, show that for the states

$$|\mathbf{p}\uparrow\rangle = e^{-(i/2)\theta\mathbf{n}\cdot\boldsymbol{\sigma}}\left|+\frac{1}{2}\right\rangle$$

$$= \cos\frac{\theta}{2}\left|+\frac{1}{2}\right\rangle + e^{i\phi}\sin\frac{\theta}{2}\left|-\frac{1}{2}\right\rangle$$

$$|\mathbf{p}\downarrow\rangle = e^{-(i/2)\theta\mathbf{n}\cdot\boldsymbol{\sigma}}\left|-\frac{1}{2}\right\rangle$$

$$= \cos\frac{\theta}{2}\left|-\frac{1}{2}\right\rangle - e^{-i\phi}\sin\frac{\theta}{2}\left|+\frac{1}{2}\right\rangle,$$

we have

$$\boldsymbol{\sigma}\cdot\mathbf{p}|\mathbf{p}\uparrow\rangle = |\mathbf{p}\uparrow\rangle$$

$$\boldsymbol{\sigma}\cdot\mathbf{p}|\mathbf{p}\downarrow\rangle = -|\mathbf{p}\downarrow\rangle.$$

$\boldsymbol{\sigma}\cdot\mathbf{p}$ is called the Helicity Operator.

Solution: Now

$$\mathbf{p} = (\sin\theta\cos\phi, \sin\theta\sin\phi, \cos\theta)$$
$$\mathbf{n} = (-\sin\phi, \cos\phi, 0)$$
$$\mathbf{n}\cdot\mathbf{p} = 0$$
$$\{\sigma_i, \sigma_j\} = 2\delta_{ij}, \quad [\sigma_i, \sigma_j] = 2i\epsilon_{ijk}\sigma_k \qquad (11.9)$$
$$\therefore \boldsymbol{\sigma}\cdot\mathbf{p}\,\boldsymbol{\sigma}\cdot\mathbf{n} = \sigma_i\sigma_j p_i n_j$$
$$= (2\delta_{ij} - \sigma_i\sigma_j)p_i n_j$$
$$= 2\mathbf{n}\cdot\mathbf{p} - \boldsymbol{\sigma}\cdot\mathbf{n}\,\boldsymbol{\sigma}\cdot\mathbf{p}$$
$$= -\boldsymbol{\sigma}\cdot\mathbf{n}\,\boldsymbol{\sigma}\cdot\mathbf{p}$$

Thus

$$\boldsymbol{\sigma}\cdot\mathbf{p}\,e^{-\frac{i}{2}\theta\mathbf{n}\cdot\boldsymbol{\sigma}} = \boldsymbol{\sigma}\cdot\mathbf{p}\left(\cos\frac{\theta}{2} - i\sin\frac{\theta}{2}\boldsymbol{\sigma}\cdot\mathbf{n}\right)$$

$$= \left(\cos\frac{\theta}{2} + i\sin\frac{\theta}{2}\boldsymbol{\sigma}\cdot\mathbf{n}\right)\boldsymbol{\sigma}\cdot\mathbf{p} \quad (11.10)$$

Now

$$\boldsymbol{\sigma}\cdot\mathbf{p} = (\sin\theta\cos\phi\,\sigma_x + \sin\theta\sin\phi\,\sigma_y + \cos\theta\,\sigma_z)$$
$$= \cos\theta\,\sigma_z + \sin\theta(\cos\phi\,\sigma_x + \sin\phi\,\sigma_y)\sigma_z\sigma_z$$

From Eq. (11.9)

$$\sigma_x\sigma_z = -i\sigma_y, \qquad \sigma_y\sigma_z = i\sigma_x$$
$$\therefore \boldsymbol{\sigma}\cdot\mathbf{p} = \cos\theta\,\sigma_z + i\sin\theta(\sin\phi\,\sigma_x - \cos\phi\,\sigma_y)\sigma_z$$
$$= (\cos\theta - i\sin\theta\,\boldsymbol{\sigma}\cdot\mathbf{n})\sigma_z \qquad (11.11)$$

and

$$\boldsymbol{\sigma}\cdot\mathbf{n}\,\boldsymbol{\sigma}\cdot\mathbf{p} = (\cos\theta\,\boldsymbol{\sigma}\cdot\mathbf{n} - i\sin\theta)\sigma_z \qquad (11.12)$$

Thus

$$\left(\cos\frac{\theta}{2} + i\sin\frac{\theta}{2}\boldsymbol{\sigma}\cdot\mathbf{n}\right)\boldsymbol{\sigma}\cdot\mathbf{p}$$

$$= \left[\cos\frac{\theta}{2}(\cos\theta - i\sin\theta\,\boldsymbol{\sigma}\cdot\mathbf{n})\right.$$

$$\left. + i\sin\frac{\theta}{2}(\cos\theta\,\boldsymbol{\sigma}\cdot\mathbf{n} - i\sin\theta)\right]\sigma_z$$

$$= \left[\left(\cos\frac{\theta}{2}\cos\theta + \sin\frac{\theta}{2}\sin\theta \right) \right.$$

$$\left. - i\boldsymbol{\sigma}\cdot\mathbf{n}\left(\cos\frac{\theta}{2}\sin\theta - \sin\frac{\theta}{2}\cos\theta \right) \right] \sigma_z$$

$$= \left(\cos\frac{\theta}{2} - i\sin\frac{\theta}{2}\boldsymbol{\sigma}\cdot\mathbf{n} \right) \sigma_z$$

$$= e^{-\frac{i}{2}\theta\boldsymbol{\sigma}\cdot\mathbf{n}}\sigma_z,$$

where we have used some trigonometrical identities. Hence from Eq. (11.10)

$$\boldsymbol{\sigma}\cdot\mathbf{p}e^{-\frac{i}{2}\theta\boldsymbol{\sigma}\cdot\mathbf{n}} = e^{-\frac{i}{2}\theta\boldsymbol{\sigma}\cdot\mathbf{n}}\sigma_z$$

Now

$$e^{-\frac{i}{2}\theta\boldsymbol{\sigma}\cdot\mathbf{n}} = \cos\frac{\theta}{2} - i\sin\frac{\theta}{2}\boldsymbol{\sigma}\cdot\mathbf{n}$$

$$= \cos\frac{\theta}{2} - i\sin\frac{\theta}{2}(-\sin\phi\sigma_x + \cos\phi\sigma_y)$$

$$= \cos\frac{\theta}{2} - i\sin\frac{\theta}{2}\left(\frac{-e^{-i\phi} - e^{i\phi}}{2i}\sigma_x + \frac{e^{i\phi} - e^{-i\phi}}{2}\sigma_y \right)$$

$$= \cos\frac{\theta}{2} + \frac{1}{2}\sin\frac{\theta}{2}[e^{i\phi}(\sigma_x - i\sigma_y) - e^{-i\phi}(\sigma_x + i\sigma_y)]$$

$$= \cos\frac{\theta}{2} + \frac{1}{2}\sin\frac{\theta}{2}(e^{-i\phi}\sigma_+ + e^{i\phi}\sigma_-)$$

Hence

$$|\mathbf{p}\uparrow\rangle = e^{-\frac{i}{2}\theta\boldsymbol{\sigma}\cdot\mathbf{n}}\left|+\frac{1}{2}\right\rangle = \cos\frac{\theta}{2}\left|+\frac{1}{2}\right\rangle + e^{i\phi}\sin\frac{\theta}{2}\left|-\frac{1}{2}\right\rangle$$

$$|\mathbf{p}\downarrow\rangle = e^{-\frac{i}{2}\theta\boldsymbol{\sigma}\cdot\mathbf{n}}\left|-\frac{1}{2}\right\rangle = \cos\frac{\theta}{2}\left|-\frac{1}{2}\right\rangle - e^{-i\phi}\sin\frac{\theta}{2}\left|+\frac{1}{2}\right\rangle$$

and

$$\boldsymbol{\sigma} \cdot \mathbf{p} |\mathbf{p} \uparrow\rangle = \boldsymbol{\sigma} \cdot \mathbf{p} e^{-\frac{i}{2}\theta\boldsymbol{\sigma}\cdot\mathbf{n}} \left|+\frac{1}{2}\right\rangle$$

$$= e^{-\frac{i}{2}\theta\boldsymbol{\sigma}\cdot\mathbf{n}} \sigma_z \left|+\frac{1}{2}\right\rangle = e^{-\frac{i}{2}\theta\boldsymbol{\sigma}\cdot\mathbf{n}} \left|+\frac{1}{2}\right\rangle = |\mathbf{p} \uparrow\rangle$$

and

$$\boldsymbol{\sigma} \cdot \mathbf{p} |\mathbf{p} \downarrow\rangle = e^{-\frac{i}{2}\theta\boldsymbol{\sigma}\cdot\mathbf{n}} \sigma_z \left|-\frac{1}{2}\right\rangle = -|\mathbf{p} \downarrow\rangle$$

Q11.16 Show that the rotation matrix for the operator $R_\omega = e^{-(i/\hbar)\boldsymbol{\omega}\cdot\mathbf{J}}$ for $\boldsymbol{\omega} = \theta\mathbf{n}$ and $\mathbf{J} = \frac{1}{2}\hbar\boldsymbol{\sigma}$ (spin $\frac{1}{2}$ operator),

$$d^{\left(\frac{1}{2}\right)}(\phi, \theta, -\phi) = \begin{pmatrix} \cos\dfrac{\theta}{2} & -e^{-i\phi}\sin\dfrac{\theta}{2} \\ e^{i\phi}\sin\dfrac{\theta}{2} & \cos\dfrac{\theta}{2} \end{pmatrix}.$$

Solution:

$$R_\omega = e^{-\frac{i}{\hbar}\boldsymbol{\omega}\cdot\mathbf{J}}$$

$$= e^{-\frac{i}{2}\theta\mathbf{n}\cdot\boldsymbol{\sigma}}$$

$$= \cos\frac{\theta}{2} - i\sin\frac{\theta}{2}\mathbf{n}\cdot\boldsymbol{\sigma}$$

Now

$$\mathbf{n} = (-\sin\phi, \cos\phi, 0)$$

$$\mathbf{n}\cdot\boldsymbol{\sigma} = -\sin\phi\sigma_x + \cos\phi\sigma_y$$

$$= \frac{i}{2}(-e^{-i\phi}\sigma_+ + e^{i\phi}\sigma_-) \tag{11.13}$$

where we have used

$$\sigma_x = \frac{1}{2}(\sigma_+ + \sigma_-)$$

$$\sigma_y = -\frac{i}{2}(\sigma_+ - \sigma_-)$$

$$\therefore R_\omega = \cos\frac{\theta}{2} + \frac{1}{2}\sin\frac{\theta}{2}(-e^{-i\phi}\sigma_+ + e^{i\phi}\sigma_-)$$

Hence

$$d_{m',m}^{1/2}(\phi, \theta) = \left\langle \frac{1}{2}m' \middle| R(\phi, \theta) \middle| \frac{1}{2}m \right\rangle$$

$$= \left\langle \frac{1}{2}m' \middle| \cos\frac{\theta}{2} + \frac{1}{2}\sin\frac{\theta}{2} \right.$$

$$\times (-e^{-i\phi}\sigma_+ + e^{i\phi}\sigma_-) \middle| \frac{1}{2}m \right\rangle$$

$$= \cos\frac{\theta}{2}\langle m'|m\rangle - e^{-i\phi}\sin\frac{\theta}{2}\langle m'|m+1\rangle$$

$$+ e^{i\phi}\sin\frac{\theta}{2}\langle m'|m-1\rangle$$

Hence

$$d^{1/2}(\phi, \theta) = \begin{pmatrix} \cos\dfrac{\theta}{2} & -e^{-i\phi}\sin\dfrac{\theta}{2} \\ e^{i\phi}\sin\dfrac{\theta}{2} & \cos\dfrac{\theta}{2} \end{pmatrix}$$

Q11.17 (a) Starting from Eq. (11.138) of the text, show that

$$\sum_{\substack{m_1 m_2 \\ m_1+m_2=m}} d_{m'_1 m_1}^{(j_1)}(\omega) d_{m'_2 m_2}^{(j_2)}(\omega) \langle j_1 j_2 m_1 m_2 | j m j_1 j_2 \rangle$$

$$= \sum_{m'=m'_1+m'_2} \langle j_1 j_2 m'_1 m'_2 | j m' j_1 j_2 \rangle d_{m'm}^{(j)}(\omega).$$

[Hint: operate Eq. (11.138) on $|j''m''j_1j_2\rangle$ and then take matrix elements with $\langle j_1j_2m_1m_2|$, use the orthogonality property of C.G. coefficients and replace at the end j'', m'' by j, m.]

(b) Use the above relation and table of C.G. coefficient to

show that

$$\sqrt{\frac{j+m}{2j}}d^{j-\frac{1}{2}}_{m'-\frac{1}{2},m-\frac{1}{2}}(\boldsymbol{\omega})d^{\frac{1}{2}}_{\frac{1}{2}\frac{1}{2}}(\boldsymbol{\omega})$$

$$+\sqrt{\frac{j-m}{2j}}d^{j-\frac{1}{2}}_{m'-\frac{1}{2},m+\frac{1}{2}}(\boldsymbol{\omega})d^{\frac{1}{2}}_{\frac{1}{2}-\frac{1}{2}}(\boldsymbol{\omega}),$$

$$=\sqrt{\frac{j+m'}{2j}}d^{j}_{m'm}(\boldsymbol{\omega}).$$

Solution: Multiply both sides of Eq. (11.138) of the text by $\langle j_1 j_2 m_1 m_2 | j'' m'' \rangle$ and sum over m_1 and m_2

$$I \equiv \sum_{\substack{m_1 m_2 \\ m_1+m_2=m''}} d^{(j_1)}_{m'_1 m_1}(\boldsymbol{\omega})d^{(j_2)}_{m'_2 m_2}(\boldsymbol{\omega})\langle j_1 j_2 m_1 m_2 | j'' m'' \rangle$$

$$= \sum_{m_1 m_2 jmm'} \langle j_1 j_2 m'_1 m'_2 | jm' j_1 j_2 \rangle \times \langle j'' m'' | j_1 j_2 m_1 m_2 \rangle$$

$$\times \langle j_1 j_2 m_1 m_2 | jm j_1 j_2 \rangle d^{(j)}_{m'm}(\boldsymbol{\omega}).$$

where we have used

$$\langle j_1 j_2 m_1 m_2 | j'' m'' \rangle$$
$$= \langle j'' m'' | j_1 j_2 m_1 m_2 \rangle, \quad \text{since C.G. are real}$$

and

$$\sum_{m_1 m_2} = |j_1 j_2 m_1 m_2 \rangle \langle j_1 j_2 m_1 m_2 | = 1$$

Then

$$I = \sum_{jmm'} \langle j_1 j_2 m'_1 m'_2 | jm' j_1 j_2 \rangle \langle j'' m'' | jm j_1 j_2 \rangle d^{(j)}_{m'm}(\boldsymbol{\omega})$$

$$= \sum_{jmm'} \delta_{j''j}\delta_{m''m}\langle j_1 j_2 m'_1 m'_2 | jm' j_1 j_2 \rangle d^{(j)}_{m'm}(\boldsymbol{\omega})$$

$$= \sum_{m'} \langle j_1 j_2 m'_1 m'_2 | j'' m' j_1 j_2 \rangle d^{(j'')}_{m'm''}(\boldsymbol{\omega})$$

Now changing $j'', m'' \to j, m$, we have

$$\sum_{\substack{m_1 m_2 \\ m_1 + m_2 = m}} d^{(j_1)}_{m'_1 m_1}(\omega) d^{(j_2)}_{m'_2 m_2}(\omega) \langle j_1 j_2 m_1 m_2 | jm \rangle$$

$$= \sum_{m' = m'_1 + m'_2} \langle j_1 j_2 m'_1 m'_2 | jm' \rangle d^{(j)}_{m'm}(\omega)$$

Take $j_2 = \frac{1}{2}$, $m_2 = \frac{1}{2}$ and $m_2 = -\frac{1}{2}$, and $m_1 = m - m_2$, therefore

$$d^{(j_1)}_{m'_1, m - \frac{1}{2}}(\omega) d^{(\frac{1}{2})}_{m'_2, \frac{1}{2}}(\omega) \left\langle j_1, \frac{1}{2}, m - \frac{1}{2}, \frac{1}{2} \middle| jm \right\rangle$$

$$+ d^{(j_1)}_{m'_1, m + \frac{1}{2}}(\omega) d^{(\frac{1}{2})}_{m'_2, -\frac{1}{2}}(\omega) \left\langle j_1, \frac{1}{2}, m + \frac{1}{2}, -\frac{1}{2} \middle| jm \right\rangle$$

$$= \sum_{m' = m'_1 + m'_2} \left\langle j_1, \frac{1}{2}, m'_1, m'_2 \middle| jm' \right\rangle d^{(j)}_{m'm}(\omega)$$

Let us take $m'_2 = \frac{1}{2}$, then $m'_1 = m' - \frac{1}{2}$ and $j = j_1 + \frac{1}{2}$ so that $j_1 = j - \frac{1}{2}$

$$\sqrt{\frac{j+m}{2j}} d^{(j - \frac{1}{2})}_{m' - \frac{1}{2}, m - \frac{1}{2}}(\omega) d^{(\frac{1}{2})}_{\frac{1}{2}, -\frac{1}{2}}(\omega)$$

$$+ \sqrt{\frac{j-m}{2j}} d^{j - \frac{1}{2}}_{m' - \frac{1}{2}, m + \frac{1}{2}}(\omega) d^{\frac{1}{2}}_{\frac{1}{2}, \frac{1}{2}}(\omega)$$

$$= \sqrt{\frac{j + m'}{2j}} d^{j}_{m'm}(\omega)$$

Q11.18 Show that

$$P_j(\cos \theta') = \frac{4\pi}{2j + 1} \sum_m Y^*_{jm}(\beta, \alpha) Y_{jm}(\theta, \phi),$$

where j is an integer; θ, ϕ and θ', ϕ' are the spherical polar coordinates of the same physical point in the old and new coordinate systems while β, α are the spherical polar coordinates of z' axis in the old coordinate system.

[Hint: consider the eigenstates $|jm\rangle$ and $|jm\rangle'$ respectively of J_z and J_z' where

$$|jm\rangle' = R_\omega|jm\rangle.$$

Take the **r**-representation and make use of $d_{m0}^{(j)}(\alpha,\beta) = \sqrt{\frac{4\pi}{2j+1}}Y_{jm}^*(\beta,\alpha)$ (c.f. Eq. (11.130)).]

Solution:

$$|jm\rangle' = R_\omega|jm\rangle$$

$$= \sum_{m'}|jm'\rangle\langle jm'|R_\omega|jm\rangle$$

$$= \sum_{m'}|jm'\rangle d_{m'm}^{(j)}(\alpha,\beta)$$

$$\langle\theta\phi|jm\rangle' = \sum_{m'}\langle\theta\phi|jm'\rangle d_{m'm}^{(j)}(\alpha,\beta)$$

$$Y_{jm}(\theta',\phi') = \sum_{m'}Y_{jm'}(\theta,\phi)d_{m'm}^{(j)}(\alpha,\beta)$$

$$Y_{j0}(\theta',\phi') = \sum_{m'}Y_{jm'}(\theta,\phi)d_{m'0}^{(j)}(\alpha,\beta)$$

$$\sqrt{\frac{2j+1}{4\pi}}P_j(\cos\theta') = \sum_{m'}Y_{jm'}(\theta,\phi)\sqrt{\frac{4\pi}{2j+1}}Y_{jm'}^*(\beta,\alpha)$$

$$P_j(\cos\theta') = \frac{4\pi}{2j+1}\sum_m Y_{jm}^*(\beta,\alpha)Y_{jm}(\theta,\phi)$$

Q11.19 Consider the 4 state system consisting of two spin-$\frac{1}{2}$ particles. The vector space of the system is spanned by the 4 orthonormal states:

$$|\uparrow\uparrow\rangle \equiv |\uparrow\rangle_1|\uparrow\rangle_2 \quad |\uparrow\downarrow\rangle \equiv |\uparrow\rangle_1|\downarrow\rangle_2$$

$$|\downarrow\downarrow\rangle \equiv |\downarrow\rangle_1|\downarrow\rangle_2 \quad |\downarrow\uparrow\rangle \equiv |\downarrow\rangle_1|\uparrow\rangle_2$$

where the arrows refer to the direction of the spin along the z-axis and the subscripts 1 and 2 refer to the particle.

Suppose that the Hamiltonian of this system is given by

$$H = \gamma(S_{1,z} + S_{2,z}) + \frac{\gamma}{\hbar}\mathbf{S}_1 \cdot \mathbf{S}_2$$

a. Write the above Hamiltonian in terms $S_{1,\pm}, S_{1,z}, S_{2,\pm}, S_{2,z}$.

b. Using the form of the Hamiltonian found in part (a) find the matrix of H in the basis given above.

c. Write the Hamiltonian in terms of the $\mathbf{S}_{total}$ where $\mathbf{S}_{total} = \mathbf{S}_1 + \mathbf{S}_2$.

d. Find the energies and stationary states of the Hamiltonian.

e. If the system is in the state $|\uparrow\downarrow\rangle$ at time $t = 0$ what is the probability of finding the system in the singlet state at time t.

Solution: (a)

$$\mathbf{S}_1 \cdot \mathbf{S}_2 = S_{1x}S_{2x} + S_{1y}S_{2y} + S_{1z}S_{2z}$$

$$S_x = \frac{S_+ + S_-}{\sqrt{2}} \qquad S_y = \frac{S_+ - S_-}{\sqrt{2i}}$$

Then

$$\mathbf{S}_1 \cdot \mathbf{S}_2 = \frac{1}{2}[S_{1+}S_{2-} + S_{1-}S_{2+}] + S_{1z}S_{2z}$$

(b) using Eq. (11.51) of the text for $j = \frac{1}{2}$, $m = \pm\frac{1}{2}$

$$S_{1+}|\uparrow\rangle = 0, \quad S_{1-}|\uparrow\rangle_1 = \hbar|\downarrow\rangle_1, S_{1z}|\uparrow\rangle = \frac{\hbar}{2}$$

$$S_{1+}|\downarrow\rangle_1 = \hbar|\uparrow\rangle, S_{1-}|\downarrow\rangle_1 = 0, S_{1z}|\downarrow\rangle = -\frac{\hbar}{2} \quad \text{etc.}$$

$$(11.14)$$

$$H|\uparrow\uparrow\rangle = \left(\gamma\hbar + \gamma\frac{\hbar}{4}\right)|\uparrow\uparrow\rangle = \frac{5}{4}\hbar|\uparrow\uparrow\rangle$$

$$H| \uparrow\downarrow\rangle = \frac{\gamma\hbar}{2}| \downarrow\uparrow\rangle - \frac{\hbar\gamma}{4}| \uparrow\downarrow\rangle$$

$$H| \downarrow\uparrow\rangle = \frac{\gamma\hbar}{2}| \uparrow\downarrow\rangle - \frac{\hbar\gamma}{4}| \downarrow\uparrow\rangle$$

$$H| \downarrow\downarrow\rangle = \left(-\gamma\hbar + \gamma\frac{\hbar}{4}\right)| \downarrow\downarrow\rangle = -\frac{3}{4}\gamma\hbar| \downarrow\downarrow\rangle$$

Contracting respectively with $| \uparrow\uparrow\rangle, \uparrow\downarrow\rangle, | \downarrow\uparrow\rangle, | \downarrow\downarrow\rangle$ and using the orthonormality of the states, it is easy to see that the matrix of H is given

$$\gamma\hbar \begin{pmatrix} \frac{5}{4} & 0 & 0 & 0 \\ 0 & -\frac{1}{4} & \frac{1}{2} & 0 \\ 0 & \frac{1}{2} & -\frac{1}{4} & 0 \\ 0 & 0 & 0 & -\frac{3}{4} \end{pmatrix}$$

Its diagonalization gives the eigenvalues as solution of the following equation

$$\left(\frac{5}{4}-\lambda\right)\left[\left(-\frac{1}{4}-\lambda\right)^2\left(-\frac{3}{4}-\lambda\right) - \frac{1}{4}(-\frac{3}{4}-\lambda)\right]$$

$$\lambda = \frac{5}{4}, \quad \lambda = -\frac{3}{4}, \quad \left[\left(\lambda+\frac{1}{4}\right)^2 - \frac{1}{4}\right] = 0$$

$$\lambda = \frac{1}{4}, -\frac{3}{4} \tag{11.15}$$

(c)

$$\mathbf{S}_{\text{total}} = \mathbf{S} = \mathbf{S}_1 + \mathbf{S}_2, \quad S_z = S_{1z} + S_{2z}$$

$$\mathbf{S}_1 \cdot \mathbf{S}_2 = \frac{1}{2}(\mathbf{S}^2 - \mathbf{S}_1^2 - \mathbf{S}_2^2)$$

$$H = \gamma S_z + \frac{\gamma}{2\hbar}(\mathbf{S}^2 - \mathbf{S}_1^2 - \mathbf{S}_2^2)$$

(d) Using Eqs. (11.76) and (11.77) of the text, we see that the simultaneous eigenstates of S_z, S^2, S_1^2, S_2^2 are the spin singlet $\frac{1}{\sqrt{2}}|\uparrow\downarrow - \downarrow\uparrow\rangle$, with eigenvalues $0, 0, \frac{3}{4}\hbar, \frac{3}{4}\hbar$ respectively, and the spin triplet $|\uparrow\uparrow\rangle, \frac{1}{\sqrt{2}}|\uparrow\downarrow + \downarrow\uparrow\rangle, |\downarrow\downarrow\rangle$ with eigenvalues $\hbar(1, 0, -1); 2\hbar, \frac{3}{4}\hbar, \frac{3}{4}\hbar$. Thus H has eigenvalues $-\frac{3\gamma}{4}\hbar, \frac{5\gamma}{4}\hbar, \frac{\gamma}{4}\hbar$ and $-\frac{3\gamma}{4}\hbar$. Hence the energies and the stationary states of H are as given above. Note that the eigenvalues are the same as obtained from the diagonalization of matrix H in (b).

(e) In the orthonormal basis given in (d), an arbitrary spin state is

$$|\chi(t)\rangle = a_0(t)|\chi_0\rangle + \sum_{i=1,2,3} a_i(t)|\chi_i\rangle$$

Then the Schrödinger equation

$$i\hbar\frac{d}{dt}|\chi(t)\rangle = H|\chi(t)\rangle$$

gives, on using the results of part (d),

$$i\hbar\frac{d}{dt}a_0(t) = E_0 a_0(t)$$

$$i\hbar\frac{d}{dt}a_i(t) = E_i a_i(t)$$

which have the solution

$$a_0(t) = e^{-iE_0t/\hbar}a_0(0), \quad E_0 = -\frac{3\gamma\hbar}{4}$$
$$a_i(t) = e^{-iE_it/\hbar}a_i(0), \quad E_i = \frac{5\gamma\hbar}{4}, \frac{\gamma\hbar}{4}, -\frac{3\gamma\hbar}{4}$$

Thus

$$|\chi(t)\rangle = e^{3i\frac{\gamma\hbar}{4}t}\left[\frac{1}{\sqrt{2}}|\uparrow\downarrow - \downarrow\uparrow\rangle\right]a_0(0)$$

$$+e^{-5i\frac{\gamma\hbar}{4}t}|\uparrow\uparrow\rangle a_1(0)$$

$$+e^{-i\frac{\gamma\hbar}{4}t}\left[\frac{1}{\sqrt{2}}|\uparrow\downarrow + \downarrow\uparrow\rangle\right]a_2(0)$$

$$+e^{i\frac{3\gamma\hbar}{4}t}|\downarrow\downarrow\rangle a_3(0)$$

Now at $t = 0$, the system is in the state $|\uparrow\downarrow\rangle$, so that

$$\tfrac{1}{\sqrt{2}}(a_0(0) + a_2(0)) = 1, \quad \tfrac{1}{\sqrt{2}}(a_0(0) - a_2(0)) = 0$$
$$a_1(0) = 0, \qquad\qquad a_3(0) = 0 \qquad (11.16)$$

Thus

$$|\chi(t)\rangle = \frac{1}{\sqrt{2}}\left[\frac{1}{\sqrt{2}}(|\uparrow\downarrow\rangle - |\downarrow\uparrow\rangle)e^{i\frac{3\gamma\hbar}{4}t}\right.$$
$$\left. + \frac{1}{\sqrt{2}}(|\uparrow\downarrow\rangle + |\downarrow\uparrow\rangle)e^{-i\frac{\gamma\hbar}{4}t}\right]$$

The probability of finding the system in spin singlet state at time t is

$$\left|\left\langle\frac{1}{\sqrt{2}}(|\uparrow\downarrow\rangle - |\downarrow\uparrow\rangle)|\chi(t)\rangle\right.\right|^2 = \left(\frac{1}{\sqrt{2}}\right)^2 = \frac{1}{2}$$

Q11.20 Consider a system of three particles. Particle 1 has spin $\frac{1}{2}$, particle 2 has spin $\frac{1}{2}$ and particle 3 has spin 1. This system has 12 states:

$$|\uparrow\uparrow\ 1\rangle\ |\uparrow\uparrow\ 0\rangle\ |\uparrow\uparrow\ -1\rangle$$
$$|\uparrow\downarrow\ 1\rangle\ |\uparrow\downarrow\ 0\rangle\ |\uparrow\downarrow\ -1\rangle$$
$$|\downarrow\uparrow\ 1\rangle\ |\downarrow\uparrow\ 0\rangle\ |\downarrow\uparrow\ -1\rangle$$
$$|\downarrow\downarrow\ 1\rangle\ |\downarrow\downarrow\ 0\rangle\ |\downarrow\downarrow\ -1\rangle$$

a. What are the possible eigenvalues of J^2_{total}?
b. For each of the eigenvalues found in part (a) what are possible eigenvalues of $J_{total,z}$.
c. Determine $(J_{1,x} + J_{2,x})^2|\uparrow\uparrow\ 1\rangle$ and $J_{1,x}J_{3,y}|\uparrow\uparrow\ 1\rangle$.
d. Write down the normalized state with total angular momentum eigenvalue 0 in terms of the individual spin states given above.

Solution: For particles 1 and 2, $j_1 = \frac{1}{2}$, $j_2 = \frac{1}{2}$ so that from Eq. (11.86) of the text, for the combined angular momentum of particles 1 and 2: $j \equiv s = 0, 1$.

Now we combine with the spin of the third particle $j_3 = 1$. Again use of Eq. (11.86) gives for $\mathbf{J}_{\text{total}} \equiv \mathbf{J}$,

$$J = 1 \qquad \text{for } j = 0$$
$$J = 0, 1, 2 \quad \text{for } j = 1$$

Thus we can write (M is the eigenvalue of J_z)

s	J	M	No of states
0	1	1,0,−1	3
	0	0	1
1	1	−1,0,1	3
	2	−2,−1,0,1,2	5

(a) The possible eigenvalues of $\mathbf{J}^2$ are $J(J+1)\hbar^2$, which in this case gives

$$2\hbar^2, 0\hbar^2, 2\hbar^2, 6\hbar^2$$

(b) Possible eigenvalues of J_z are as given in the above table
(c)

$$(J_{1,x} + J_{2,x})^2 \equiv (s_{1,x} + s_{2,x})^2$$
$$= \frac{1}{2}[(s_{1+} + s_{1-}) + (s_{2+} + s_{2-})]^2$$

Acting on $|\uparrow\uparrow 1\rangle$, the only terms which give non zero values are

$$\frac{1}{2}[(s_{1+}s_{1-} + s_{2+}s_{2-})$$
$$+ (s_{1-}s_{2-} + s_{2-}s_{1-})]|\uparrow\uparrow 1\rangle$$
$$= \frac{1}{2}[|\uparrow\uparrow 1\rangle + |\uparrow\uparrow 1\rangle + |\downarrow\downarrow 1\rangle + |\downarrow\downarrow 1\rangle]$$
$$= |\uparrow\uparrow 1\rangle + |\downarrow\downarrow 1\rangle$$

$$J_{1x}J_{3y}|\uparrow\uparrow 1\rangle = \frac{1}{\sqrt{2}}(s_{1+}+s_{1-})\frac{1}{\sqrt{2}i}(J_{3+}-J_{3-})|\uparrow\uparrow 1\rangle$$

$$= \frac{1}{2i}[-(s_{1+}+s_{1-})\sqrt{2}\hbar]|\uparrow\uparrow 0\rangle,$$

where we have used

$$J_{3+}|1\rangle = 0 \quad J_{3-}|1\rangle = \sqrt{2}\hbar|0\rangle$$

$$s_{1+}|\uparrow\rangle = 0 \quad s_{1-}|\uparrow\rangle = \hbar|\downarrow\rangle$$

(d) It is clear from the table above that the state with total angular momentum eigenvalue involves spin triplet and in order to ensure that J_z has eigenvalue 0, the only possible state is a linear combination of

$$|\uparrow\uparrow -1\rangle, \quad \frac{1}{\sqrt{2}}(|\uparrow\downarrow\rangle + |\downarrow\uparrow 0\rangle), \quad |\downarrow\downarrow 1\rangle$$

So the required normalized state is

$$\frac{1}{\sqrt{3}}\left[|\uparrow\uparrow -1\rangle + \frac{1}{\sqrt{2}}(|\uparrow\downarrow\rangle + |\downarrow\uparrow 0\rangle) + |\downarrow\downarrow 1\rangle\right]$$

Q11.21 A particle of spin $\frac{1}{2}$ is in a D-state of the orbital angular momentum:

a. What are its possible states of total angular momentum?

b. For the Hamiltonian

$$H = a + b\mathbf{L}\cdot\mathbf{S} + c\mathbf{L}^2$$

where a, b and c are numbers. Find the values of the energy for each of the different states of total angular momentum (express your answer in terms of a, b, c).

$$j = \ell + \frac{1}{2}, \quad \ell - \frac{1}{2}$$

Thus for D-state $\ell = 2$;

$$j = \frac{5}{2}, \frac{3}{2} : \quad D_{5/2}, \quad D_{3/2}$$

Solution: Now (c.f. Eq. (11.108) of the text)

$$\mathbf{J} = \mathbf{L} + \mathbf{S}$$

$$\therefore \mathbf{L} \cdot \mathbf{S} = \frac{1}{2}(J^2 - L^2 - S^2)$$

Thus the eigenvalues of $\mathbf{L} \cdot \mathbf{S}$ are:

$$\langle \mathbf{L} \cdot \mathbf{S} \rangle = \frac{1}{2}\hbar^2[j(j+1) - \ell(\ell+1) - s(s+1)]$$

$$= \frac{\hbar^2}{2}\left[j(j+1) - 6 - \frac{3}{4}\right]$$

$$\therefore \langle \mathbf{L} \cdot \mathbf{S} \rangle_{j=5/2} = \hbar^2, \quad \langle \mathbf{L} \cdot \mathbf{S} \rangle_{j=3/2} = -\hbar^2$$

Hence the eigenvalues of H for $D_{5/2}$ and $D_{3/2}$ states are:

$$E_{5/2} = a + b\hbar^2 + 6c\hbar^2$$

$$E_{3/2} = a - b\hbar^2 + 6c\hbar^2$$

Q11.22 Show that:

(a)

$$e^{\frac{i}{\hbar}\frac{\pi}{2}J_y} J_x e^{\frac{-i}{\hbar}\frac{\pi}{2}J_y} = J_z$$

(b)

$$e^{\frac{i}{\hbar}\frac{\pi}{2}J_y} e^{\frac{i}{\hbar}\alpha J_x} e^{\frac{-i}{\hbar}\frac{\pi}{2}J_y} = e^{\frac{i}{\hbar}\alpha J_z}$$

(c) For any vector operators $\mathbf{A}$:

$$e^{\frac{i}{\hbar}\alpha J_y} A_x e^{\frac{-i}{\hbar}\alpha J_y} = A_x \cos\alpha + A_z \sin\alpha$$

Solution: Now

$$e^{i\frac{\alpha}{\hbar}J_y} J_x e^{-i\frac{\alpha}{\hbar}J_y} = J_x - i\alpha[J_x, J_y] - \frac{\alpha^2}{2}i\hbar[J_z, J_y] + \cdots$$

$$= J_x + \alpha J_z - \frac{\alpha^2}{2}J_x + \cdots$$

$$= J_x\left(1 - \frac{\alpha^2}{2} + \cdots\right) + J_z\left(\alpha - \frac{\alpha^2}{2!} + \cdots\right)$$

$$= J_x \cos\alpha + J_z \sin\alpha \qquad (11.17)$$

(a) For $\alpha = \frac{\pi}{2}$

$$e^{i\frac{\pi}{2\hbar}J_y} J_x e^{-i\frac{\pi}{2\hbar}J_y} = J_z$$

(b) Now

$$e^{i\frac{\pi}{2\hbar}J_y} e^{i\frac{\alpha}{\hbar}J_x} e^{-i\frac{\pi}{2\hbar}J_y}$$

$$= e^{i\frac{\pi}{2\hbar}J_y} \left(1 + i\frac{\alpha}{\hbar}J_x + \left(i\frac{\alpha}{\hbar}\right)^2 J_x^2 + \cdots\right) e^{-i\frac{\pi}{2\hbar}J_y}$$

$$= 1 + i\frac{\alpha}{\hbar}J_z - \left(\frac{\alpha}{\hbar}\right)^2 J_z^2 + \cdots = e^{i\frac{\alpha}{\hbar}J_z}$$

(c) Since for any vector operator $\mathbf{A}$,

$$[J_i, A_i] = i\hbar\epsilon_{ijk}A_k$$

Hence from Eq. (11.17)

$$e^{i\frac{\alpha}{\hbar}J_y} A_x e^{-i\frac{\alpha}{\hbar}J_y} = A_x \cos\alpha + A_z \sin\alpha$$

Q11.23 Show that:

(a)

$$[J_x, [J_x, T_q^{(k)}]] = \sum_{q'} T_{q'}^{(k)} \langle kq'|J_x^2|kq\rangle$$

(b)

$$[J_x, [J_x, T_q^{(k)}]] + [J_y, [J_y, T_q^{(k)}]] + [J_z, [J_z, T_q^{(k)}]]$$
$$= k(k+1)\hbar^2 T_q^{(k)}$$

Solution: From Eq. (11.151c) of the text,

$$[\mathbf{J}, T_q^{(k)}] = \sum_{q'} T_{q'}^{(k)} \langle kq'|\mathbf{J}|kq\rangle$$

we have

(a)

$$[J_x, [J_x, T_q^{(k)}]] = \sum_{q''} \langle kq''|J_x|kq\rangle [J_x, T_{q'}^{(k)}]$$

$$= \sum_{q''} \sum_{q'} \langle kq''|J_x|kq\rangle T_{q'}^{(k)} \langle kq'|J_x|qk''\rangle$$

$$= \sum_{q''} \sum_{q'} T_{q'}^{(k)} \langle kq'|J_x|qk''\rangle \langle kq''|J_x|kq\rangle$$

$$= \sum_{q'} T_{q'}^{(k)} \langle kq'|J_x^2|kq\rangle$$

(b)

$$[J_x, [J_x, T_q^{(k)}]] + [J_y, [J_y, T_q^{(k)}]] + [J_z, [J_z, T_q^{(k)}]]$$

$$= \sum_{q'} T_{q'}^{(k)} \langle kq'|(J_x^2 + J_y^2 + J_z^2)|kq\rangle$$

$$= \sum_{q'} T_{q'}^{(k)} \langle kq'|J^2|kq\rangle$$

$$= \sum_{q'} T_{q'}^{(k)} k(k+1)\hbar^2 \delta_{q'q} = k(k+1)\hbar^2 T_q^{(k)}$$

Additional Problems:

Q1 (a) Show that for a vector $\mathbf{A}$,

(i)

$$\mathbf{J}.\mathbf{A} = \mathbf{A}.\mathbf{J}$$

(ii)

$$[J^2, \mathbf{A}] = -i\hbar[\mathbf{J} \times \mathbf{A} - \mathbf{A} \times \mathbf{J}]$$

(iii)

$$[J^2, [J^2, \mathbf{A}]] = 2\hbar^2[J^2\mathbf{A}+\mathbf{A}J^2] - 2\hbar^2[\mathbf{J}(\mathbf{J}.\mathbf{A})+(\mathbf{J}.\mathbf{A})\mathbf{J}]$$

(iv)

$$\langle jm'|\mathbf{A}|jm\rangle = \frac{1}{2j(j+1)\hbar^2}$$
$$\times [\langle jm'|[(\mathbf{J}.\mathbf{A})\mathbf{J} + \mathbf{J}(\mathbf{J}.\mathbf{A})]|jm\rangle]$$

(v)

$$\langle jm|A_z|jm\rangle = \frac{m}{j(j+1)\hbar}\langle jm|\mathbf{J}.\mathbf{A}|jm\rangle$$

In particular for the magnetic moment $\boldsymbol{\mu}$

$$\langle jm|\mu_z|jm\rangle = \frac{m}{j(j+1)\hbar}\langle jm|\boldsymbol{\mu}.\mathbf{J}|jm\rangle$$

(b) If

$$\boldsymbol{\mu} = \mu_1\mathbf{J}_1 + \mu_2\mathbf{J}_2$$
$$\mathbf{J} = \mathbf{J}_1 + \mathbf{J}_2,$$

Show that

$$\mu = \langle jj|\mu_z|jj\rangle$$
$$= \frac{\hbar}{2(j+1)\hbar}\{(\mu_1 + \mu_2)j(j+1)$$
$$+ (\mu_1 - \mu_2)[j_1(j_1+1) - j_2(j_2+1)]\}$$

Solution: (a)

(i)

$$[J_i, A_j] = i\epsilon_{ijk}A_k$$
$$J_iA_j = A_jJ_i + i\epsilon_{ijk}A_k$$

For $j = i$

$$J_iA_i = A_iJ_i$$
$$\mathbf{J}.\mathbf{A} = \mathbf{A}.\mathbf{J}$$

(ii)

$$[\mathbf{J}^2, A_j] = [J_iJ_i, A_j]$$
$$= J_i[J_i, A_j] + [J_i, A_j]J_i$$
$$= i\hbar\epsilon_{ijk}J_iA_k + i\hbar\epsilon_{ijk}A_kJ_i$$
$$= -i\hbar\epsilon_{jik}J_iA_k + i\hbar\epsilon_{jki}A_kJ_i$$
$$= -i\hbar[(\mathbf{J}\times\mathbf{A})_j - (\mathbf{A}\times\mathbf{J})_j] \quad (11.18)$$

Hence

$$[\mathbf{J}^2, \mathbf{A}] = -i\hbar[(\mathbf{J}\times\mathbf{A}) - (\mathbf{A}\times\mathbf{J})]$$

(iii) On using Eq. (11.18)

$$[\mathbf{J}^2, [\mathbf{J}^2, A_j]] = i\hbar\epsilon_{ijk}[\mathbf{J}^2, (J_i A_k + A_k J_i)]$$
$$= i\hbar\epsilon_{ijk}\{[\mathbf{J}^2, J_i A_k] + [\mathbf{J}^2, A_k J_i]\}$$

Now $[\mathbf{J}^2, J_i] = 0$.
Thus

$$[\mathbf{J}^2, [\mathbf{J}^2, A_j]] = i\hbar\epsilon_{ijk}\{J_i[\mathbf{J}^2, A_k] + [\mathbf{J}^2, A_k]J_i\}$$
$$= -\hbar^2\epsilon_{ijk}\epsilon_{nkr}\{J_i(J_n A_r + A_r J_n)$$
$$+ (J_n A_r + A_r J_n)J_i\}$$
$$= \hbar^2(\delta_{in}\delta_{jr} - \delta_{ir}\delta_{nj})\{J_i(J_n A_r + A_r J_n)$$
$$+ (J_n A_r + A_r J_n)J_i\}$$
$$= \hbar^2[\mathbf{J}^2 A_j + A_j\mathbf{J}^2 - (\mathbf{J.A})J_j - J_j(\mathbf{J.A})]$$
$$+ J_i(A_j J_i - J_i A_i) + (J_i A_j - A_i J_j)J_i$$

Now

$$J_i(A_j J_i - J_j A_i) + (J_i A_j - A_i J_j)J_i$$
$$= J_i[J_i A_j - i\hbar\epsilon_{ijn}A_n - A_i J_j - i\hbar\epsilon_{jin}A_n]$$
$$+ [A_j J_i + i\hbar\epsilon_{ijn}A_n - J_j A_i + i\hbar\epsilon_{jin}A_n]J_i$$
$$= \mathbf{J}^2 A_j - \mathbf{J.A}J_j + A_j J^2 - J_j\mathbf{A.J}$$
$$= \mathbf{J}^2 A_j - \mathbf{J.A}J_j + A_j\mathbf{J}^2 - J_j(\mathbf{J.A})$$

Hence

$$[\mathbf{J}^2, [\mathbf{J}^2, \mathbf{A}]] = 2\hbar^2[\mathbf{J}^2\mathbf{A} + \mathbf{A}\mathbf{J}^2 - (\mathbf{J.A})\mathbf{J} - \mathbf{J}(\mathbf{J.A})]$$
$$(11.19)$$

(iv)

$$\langle jm'|[\mathbf{J}^2, [\mathbf{J}^2, \mathbf{A}]]|jm\rangle = j(j+1)[\langle jm'|[J^2, \mathbf{A}]|jm\rangle - \text{same}]$$
$$(11.20)$$
$$= 0$$

From Eqs. (11.19) and (11.20)

$$2\hbar^2 \langle jm'|[\mathbf{J}^2\mathbf{A}+\mathbf{A}\mathbf{J}^2-(\mathbf{J}.\mathbf{A})\mathbf{J}-\mathbf{J}(\mathbf{J}.\mathbf{A})]j_m\rangle = 0 \qquad (11.21)$$

or

$$\langle jm'|\mathbf{A}|jm\rangle = \frac{1}{2j(j+1)\hbar^2}[\langle jm'|[(\mathbf{J}.\mathbf{A})\mathbf{J} + \mathbf{J}(\mathbf{J}.\mathbf{A})]|jm\rangle]$$

(v)

$$\langle jm'|A_z|jm\rangle = \frac{2m\hbar\delta_{m'm}}{2j(j+1)\hbar^2}\langle jm'|\mathbf{J}.\mathbf{A}|jm\rangle$$

Hence

$$\langle jm|A_z|jm\rangle = \frac{m}{j(j+1)\hbar}\langle jm|\mathbf{J}.\mathbf{A}|jm\rangle \qquad (11.22)$$

$$\langle jm|\mu_z|jm\rangle = \frac{m}{j(j+1)\hbar}\langle jm|\boldsymbol{\mu}.\mathbf{J}|jm\rangle \qquad (11.23)$$

(b)

$$\boldsymbol{\mu}.\mathbf{J} = \mu_1\mathbf{J}_1.\mathbf{J} + \mu_2\mathbf{J}_2.\mathbf{J}$$
$$= \mu_1\mathbf{J}_1^2 + \mu_2\mathbf{J}_2^2 + (\mu_1 + \mu_2)\mathbf{J}_1.\mathbf{J}_2$$

Now

$$\mathbf{J}_1.\mathbf{J}_2 = \frac{1}{2}(\mathbf{J}^2 - \mathbf{J}_1^2 - \mathbf{J}_2^2)$$

$$\boldsymbol{\mu}.\mathbf{J} = \frac{1}{2}(\mu_1 - \mu_2)(\mathbf{J}_1^2 - \mathbf{J}_2^2) + \frac{1}{2}(\mu_1 + \mu_2)\mathbf{J}^2$$
$$\qquad (11.24)$$

$$|jm\rangle = \sum_{m_1m_2} |j_1j_2m_1m_2\rangle\langle j_1j_2m_1m_2|jm\rangle$$

$$J^2|jm\rangle = j(j+1)\hbar^2|jm\rangle$$

$$J_1^2|jm\rangle = j_1(j_1+1)\hbar^2|jm\rangle$$

$$J_2^2|jm\rangle = j_2(j_2+1)\hbar^2|jm\rangle \qquad (11.25)$$

Hence from Eqs. (11.23), (11.24) and (11.25), the magnetic
moment

$$\mu \equiv \langle jj|\mu_z|jj\rangle = \frac{\hbar}{2(j+1)}\{(\mu_1 + \mu_2)j(j+1)$$

$$+(\mu_1 - \mu_2)(j_1(j_1+1) - j_2(j_2+1))\} \qquad (11.26)$$

Chapter 12

Time Independent Perturbation Theory

Q12.1 Consider a harmonic oscillator of mass m on which acts a uniform time-independent external force F. Call its ground state

$$\tilde{u}_0(x) = \langle x|\tilde{0}\rangle$$

If

$$u_1(x) = \langle x|1\rangle$$

is the first excited state of the same oscillator when $F = 0$, show that to the first order in perturbation the probability of finding it in the first excited state is

$$|\langle 1|\tilde{0}\rangle|^2 = \left|\int u_1^*(x)\tilde{u}_0(x)dx\right|^2$$

$$= \frac{F^2}{2m\hbar}\frac{1}{\omega^3}.$$

Also calculate the shift in the ground state energy to second order in F.

Solution:

$$F = -\frac{\partial V}{\partial x}$$

$$\lambda V(x) = -Fx$$

Thus

$$H = H_0 - Fx$$

Now as given

$$H\tilde{u}_0(x) = E_0\tilde{u}_0(x)$$

But

$$H_0 u_1(x) = \epsilon_1 u_1(x)$$

Thus

$$\tilde{u}_0(x) = u_0 - F \sum_{n \neq 0} \frac{\langle n|x|0\rangle u_n}{\epsilon_0 - \epsilon_n}$$

Now

$$\langle n|x|0\rangle = \delta_{n1}\sqrt{\frac{\hbar}{2m\omega}} \quad \text{[From problem 9.5]} \quad (12.1)$$

Hence

$$\tilde{u}_0 = u_0 - F\sqrt{\frac{\hbar}{2m\omega}}\frac{1}{(-\hbar\omega)}u_1,$$

$$\langle 1|\tilde{0}\rangle = \int u_1^*(x)\tilde{u}_0(x)dx$$

$$= \int u_1^* \left[u_0 + \sqrt{\frac{\hbar}{2m\omega}}\frac{1}{\hbar\omega}u_1 \right] dx$$

$$= F\sqrt{\frac{\hbar}{2m\omega}}\frac{1}{\hbar\omega}$$

and

$$|\langle 1|\tilde{0}\rangle|^2 = \frac{F^2}{2m\hbar\omega^3}$$

Q12.2 A hydrogen atom is placed in an external constant field F in the z-direction. For the ground state s of the atom, write

$$e^2 F^2 \sum_{n \neq s} \frac{|V_{sn}|^2}{\varepsilon_s - \varepsilon_n} = -\frac{1}{2} F^2 \alpha$$

Show that

$$4a^3 < \alpha < \frac{16}{3} a^3$$

where a is the Bohr radius $\frac{\hbar^2}{me^2}$.

Solution: Perturbation potential is

$$eV = e\mathbf{F} \cdot \mathbf{r} = Fz$$

$$\Delta E = -e^2 F^2 \sum_{n>1} \frac{|\langle 1S|z|n\ell m\rangle|^2}{(\epsilon_n - \epsilon_1)},$$

where $|n\ell m\rangle$ is an eigenstate of the hydrogen atom with principal quantum number n, angular momentum ℓ, and $\ell_z = m$. The energy levels of the hydrogen atom, ϵ_n, are given by $\epsilon_n = -\frac{e^2}{2n^2 a^2}$. Since $|\langle 1S|z|n\ell m\rangle|^2$ is always positive, and $-\frac{1}{\epsilon_1} < \frac{1}{\epsilon_n - \epsilon_1} < \frac{1}{\epsilon_2 - \epsilon_1}$, it follows that

$$-\frac{e^2}{\epsilon_1} \sum |\langle 1S|z|n\ell m\rangle|^2 < \frac{\alpha}{2} < \frac{e^2}{(\epsilon_2 - \epsilon_1)} \sum |\langle 1S|z|n\ell m\rangle|^2$$

In setting the lower limit, the contribution from the continuum states is assumed negligible. Since the states $|n\ell m\rangle$ form a complete set (neglecting the continuum),

$$\sum |\langle 1S|z|n\ell m\rangle\langle n\ell m|z|1S\rangle| = \langle 1S|z^2|1S\rangle$$

$$= \langle 1S|r^2 \cos^2 \theta|1S\rangle$$

Now

$$|1S\rangle = R_{10}Y_{00}$$

Hence

$$\langle 1S|r^2\cos^2\theta|1S\rangle = \langle r^2\rangle_{10}\left[\int\cos^2\theta|Y_{00}|^2 2\pi\sin\theta d\theta\right]$$

$$= \frac{1}{3}\langle r\rangle_{10}$$

$$= a^2 \quad \text{[c.f. page 114 of the text]}$$

Writing $\epsilon_1 = -e^2/2a$ and $\epsilon_2 - \epsilon_1 = 3e^2/8a$, one finally obtains $4a^2 < \alpha < (16/3)a^3$. The experimental value is $\alpha = 4.5a^3$.

Q12.3 If the perturbation potential is given by

$$eV = e\mathbf{F}\cdot\mathbf{r} = eFr\cos\theta$$

where F is a constant, show that for the hydrogen atom

$$\langle 100|V|nlm\rangle = 0 \text{ for } m \neq 0$$

and $\langle 100|V|nl0\rangle$ is non-zero only for $l = 1$ and in this case it is given by

$$eF\frac{1}{\sqrt{3}}\int_0^\infty R_{10}^*(r)R_{n1}(r)r^3 dr$$

Hence show that for the ground state

$$E_1 = \epsilon_1 + \frac{e^2 F^2}{3}\sum_{n\neq 1}\frac{|\int_0^\infty R_{10}R_{n1}r^3 dr|^2}{\epsilon_1 - \epsilon_n}$$

Solution:

$$\langle 100|r\cos\theta|n\ell n\rangle = \int u_{100}^* r\cos\theta\, u_{n\ell m} d\mathbf{r}$$

$$= \int R_{10} r R_{n\ell} r^2 dr \int \cos\theta\, Y_{00}^* Y_{\ell m} d\Omega$$

Now as shown in the previous problem

$$\int \cos\theta\, Y_{00}^* Y_{\ell m} d\Omega = \frac{1}{\sqrt{3}}\delta_{\ell 1}\delta_{m0}$$

Thus only $\ell = 1$ and $m = 0$ contribute. Hence

$$\langle 100|r\cos\theta|n10\rangle = \frac{1}{\sqrt{3}}\int R_{10} r R_{n1} r^2 dr$$

and

$$\langle 100|r\cos\theta|100\rangle = 0$$

Hence 1st order perturbation gives zero and 2nd order perturbation gives

$$E_1 = \epsilon_1 + \frac{e^2 F^2}{3}\sum_{n\neq 1}\frac{|\int_0^\infty R_{10}R_{n1}r^3 dr|^2}{\epsilon_1 - \epsilon_n}$$

Q12.4 Consider an atom which has a nucleus of charge Z and one electron. Using first order perturbation theory, calculate the energy shift of the $1s$, $2s$ and $2p$ states of ^{1}H and ^{235}U atoms, assuming that the nucleus is a uniformly charged sphere of radius $R = r_0 A^{1/3}$, where $r_0 \approx 1.2 \times 10^{-13}$ cm and A is the atomic number. Note that $R \ll a$, where a is the Bohr radius.

Hint: The Coulomb potential is given by

$$V(r) = -\frac{3Ze^2}{2R^3}\left(R^2 - \frac{1}{3}r^2\right) \quad r < R$$

$$= -\frac{Ze^2}{r} \quad r > R.$$

Solution: We can write

$$V(r) = -\frac{Ze^2}{r} + \lambda v(r)$$

where

$$\lambda v(r) = \begin{cases} 0 & r > R \\ (Ze^2)\left[\dfrac{1}{r} - \dfrac{3}{2R} + \dfrac{1}{2}\dfrac{r^2}{R^3}\right] & r < R \end{cases}$$

1st order perturbation theory gives

$$E_{n\ell} = \epsilon_n + \lambda\langle V\rangle_{n\ell}$$

$$\lambda\langle V\rangle_{n\ell} = \lambda \int u_{n\ell m}^* v(r) v_{n\ell m} d^3r$$

$$= \lambda\langle n\ell m|v|n\ell m\rangle$$

and

$$\int_0^\pi \int_0^{2\pi} Y_{10}^* Y_{10} d\Omega = 1$$

Since perturbation does not involve θ and ϕ and therefore,

$$\langle n\ell m|v|n\ell'm'\rangle = \int_0^\infty \int_0^{2\pi} \int_0^\pi u_{n\ell m}^* v u_{n\ell'm'} r^2 dr d\Omega$$

$$= \int_0^\infty R_{n\ell} v(r) R_{n\ell'} r^2 dr \int Y_{\ell m}^* Y_{\ell'm'} d\Omega$$

where

$$\int Y_{\ell m}^* Y_{\ell'm'} d\Omega = \delta_{ll'}\delta_{mm'}$$

Thus we need to consider

$$\langle n\ell m|v|n\ell m\rangle = \int_0^R R_{n\ell}^* v R_{n\ell} r^2 dr \quad \text{since}$$

$$v = 0 \text{ for } r > R$$

For $n = 1$:

$$\langle 100|v|100 \rangle = \int_0^R R_{10}^* v(r) R_{10} r^2 dr \qquad (12.2)$$

giving

$$\int_0^R R_{10} \left(\frac{3}{2R} - \frac{1}{2}\frac{r^2}{R^3} - \frac{1}{r} \right) R_{10} r^2 dr$$

where

$$R_{10} = 2 \left(\frac{Z}{a_0} \right)^{3/2} e^{-Zr/a_0} \quad 0 < r < \infty$$

$$\approx 2 \left(\frac{Z}{a_0} \right)^{3/2}, \quad \text{for } \frac{R}{a_0} \ll 1 \qquad (12.3)$$

Thus

$$\langle 100|v|100 \rangle = 4 \left(\frac{Z}{a_0} \right)^3 \int_0^R \left(\frac{3}{2R} - \frac{1}{2}\frac{r^2}{R^3} - \frac{1}{r} \right) r^2 dr$$

$$= -\frac{2}{5}\frac{Z^3}{a_0} \left(\frac{R}{a_0} \right)^2 \qquad (12.4)$$

Hence

$$E_{1s} = -\frac{Z^2 e^2}{2a_0} - Ze^2 \left(\frac{-4}{10} \right) \frac{Z^3}{a_0} \left(\frac{R}{a_0} \right)^2$$

$$= -\frac{Z^2 e^2}{2a_0} \left[1 - \frac{4}{5} \left(\frac{r_0}{a_0} \right)^2 Z^2 A^{2/3} \right] \qquad (12.5)$$

For $n = 2$:

$$2s: \quad \langle 200|v|200 \rangle = \int_0^R R_{20}^* \left(\frac{3}{2R} - \frac{1}{2}\frac{r^2}{R^3} - \frac{1}{r} \right) R_{20} r^2 dr$$

$$2p: \langle 21m|v|21m \rangle = \int_0^R R_{21}^* \left(\frac{3}{2R} - \frac{1}{2}\frac{r^2}{R^3} - \frac{1}{r} \right) R_{21} r^2 dr$$

where $m = 1, 0, -1$

Now

$$R_{20} = \left(\frac{Z}{2a_0}\right)^{3/2}\left(2 - \frac{Zr}{a_0}\right)e^{-Zr/2a_0}$$

$$R_{21} = \left(\frac{Z}{2a_0}\right)^{3/2}\frac{Z}{\sqrt{3}}\frac{r}{a_0}e^{-Zr/2a_0}$$

Now for the ground states and the first excited state:

$$1s: \qquad u_{n\ell m} = u_{100} \qquad \epsilon_{1s} = -\frac{Ze^2}{2a_0}$$

$$2s: \qquad u_{n\ell m} = u_{200}$$

$$2p: \qquad u_{n\ell m} = u_{21m} \qquad m = -1, 0, 1$$

$$\epsilon_{2s} = \epsilon_{2p} = -\frac{Ze^2}{2a_0}\frac{1}{4}$$

Thus the $n = 2$ level is 4-fold degenerate.
Since

$$R \ll a_0, \qquad \text{we write}$$

$$R_{20} \approx 2\left(\frac{Z}{2a_0}\right)^{3/2}, \qquad R_{21} \simeq \left(\frac{Z}{2a_0}\right)^{3/2}\frac{Zr}{\sqrt{3}a_0}$$

$\therefore$ for $n = 2$, we get

$$\langle 200|v|200\rangle \approx 4\left(\frac{Z}{2a_0}\right)^3\int_0^R\left(\frac{3}{2R} - \frac{1}{2}\frac{r^2}{R^3} - \frac{1}{r}\right)r^2dr$$

$$= -\frac{1}{20}\frac{Z}{a_0}\left(\frac{ZR}{a_0}\right)^2 = A$$

$$\langle 21m|v|21m\rangle \approx \left(\frac{Z}{2a_0}\right)^3\frac{Z^2}{3a_0^2}\int_0^R r^2\left(\frac{3}{2R} - \frac{1}{2}\frac{r^2}{R^3} - \frac{1}{r}\right)r^2dr$$

$$= -\left(\frac{1}{140}\right)\frac{Z}{8a_0}\left(\frac{ZR}{a_0}\right)^4 = B \qquad m = 1, 0, -1$$

Hence for $n = 2$

$$\epsilon_{2s} = -\frac{Ze^2}{8a_0^2} = \epsilon_{2p} \equiv \epsilon_2$$

$$E_{2s} = \epsilon_2 - Ze^2 A$$

$$= -\frac{Z^2 e^2}{8a_0} \left[1 - \frac{4}{5} \left(\frac{ZR}{a_0} \right)^2 \right]$$

$$E_{2p} = \epsilon_2 - Ze^2 B$$

$$= -\frac{Z^2 e^2}{8a_0} \left[1 - \underbrace{\frac{1}{140} \left(\frac{ZR}{a_0} \right)^4}_{\text{negligible compared to } \frac{4}{5} \left(\frac{ZR}{a_0} \right)^2} \right]$$

Note that $2s$ and $2p$ levels, which were degenerate, now split due to finite size of the nucleus

$$E_{2s} - E_{2p} \approx \frac{Z^2 e^2}{8a_0} \frac{4}{5} \left(\frac{ZR}{a_0} \right)^2$$

However this splitting is extremely small particularly for lighter atoms.

$2p$ levels with respect to the magnetic quantum number m are still degenerate.

Q12.5 **Zeeman Effect:**

The Hamiltonian for a hydrogen like atom can be written as

$$H_0 = \frac{\hat{\mathbf{p}}^2}{2m_e} - \frac{Ze^2}{r} + \frac{Ze^2}{2m_e^2 c^2 r^3} \mathbf{S} \cdot \mathbf{L}$$

The last term represents the interaction of the magnetic moment of the electron $\boldsymbol{\mu} = -\frac{e}{m_e c} \mathbf{S}$ with the magnetic field $\mathbf{B} = \frac{1}{2c} \mathbf{E} \times \mathbf{v} = \frac{1}{e} \frac{Ze}{r^3} \mathbf{r} \times \mathbf{v} = \frac{1}{2m_e c} \frac{Ze}{r^3} \mathbf{r} \times \mathbf{p}$ which arises due to the motion of an electron in the electric field E of nucleus of charge Ze.

Consider the atom in an external weak homogeneous magnetic field B along the z-axis so that the magnetic interaction energy is given by

$$V = -\boldsymbol{\mu}_L \cdot \mathbf{B} - \boldsymbol{\mu} \cdot \mathbf{B} = \frac{e}{2m_e c}BL_z + \frac{e}{m_e c}BS_z$$

$$= \frac{eB}{2m_e c}(L_z + 2S_z).$$

It is clear that the eigenstates of H_0 are $|njm\rangle$,

$$H_0|njm\rangle = E_{nj}|njm\rangle$$

Treating V as a perturbation, and using Eqs. (11.95) and (11.96), show that the energy shift

$$\Delta E_{nl}^B = \langle njm|V|njm\rangle$$

$$= \frac{eB}{2m_e c}\langle njm|(L_z + 2S_z)|njm\rangle$$

$$= \frac{eB\hbar}{2m_e c}m\left(1 \pm \frac{1}{2l+1}\right), \quad j = l \pm 1/2.$$

Solution:

$$\mathbf{J} = \mathbf{L} + \mathbf{S}$$

$$J_z = L_z + S_z$$

. Now

$$H_0|njm\rangle = E_{nj}|njm\rangle,$$

where

$$j = \ell + \frac{1}{2}, \quad \ell - \frac{1}{2}$$

From Eqs. (11.91) and (11.92) of the text

$$\left| n, \ell + \frac{1}{2} m \right\rangle = \sqrt{\frac{\ell + m + \frac{1}{2}}{2\ell + 1}} \left| n, \ell \frac{1}{2} m - \frac{1}{2} \frac{1}{2} \right\rangle$$

$$+ \sqrt{\frac{\ell - m + \frac{1}{2}}{2\ell + 1}} \left| n, \ell \frac{1}{2} m + \frac{1}{2} - \frac{1}{2} \right\rangle \quad (12.6)$$

Noting that

$$\left| \ell \frac{1}{2} m - \frac{1}{2} \frac{1}{2} \right\rangle = \left| \ell m - \frac{1}{2} \right\rangle \left| \frac{1}{2} \frac{1}{2} \right\rangle$$

$$\left| \ell \frac{1}{2} m + \frac{1}{2} - \frac{1}{2} \right\rangle = \left| \ell m + \frac{1}{2} \right\rangle \left| \frac{1}{2} - \frac{1}{2} \right\rangle,$$

we get from Eq. (12.6)

$$(L_z + 2S_z) \left| \ell + \frac{1}{2} m \right\rangle$$

$$= \sqrt{\frac{\ell + m + \frac{1}{2}}{2\ell + 1}} \left(m + \frac{1}{2} \right) \hbar \left| \ell \frac{1}{2} m - \frac{1}{2} \frac{1}{2} \right\rangle$$

$$+ \sqrt{\frac{\ell - m + \frac{1}{2}}{2\ell + 1}} \left(m - \frac{1}{2} \right) \hbar \left| \ell \frac{1}{2} m + \frac{1}{2} - \frac{1}{2} \right\rangle$$

and

$$(L_z + 2S_z) \left| l - \frac{1}{2} m \right\rangle$$

$$= -\sqrt{\frac{\ell - m + \frac{1}{2}}{2\ell + 1}} \left(m + \frac{1}{2} \right) \hbar \left| \ell \frac{1}{2} m - \frac{1}{2} \frac{1}{2} \right\rangle$$

$$+ \sqrt{\frac{\ell + m + \frac{1}{2}}{2\ell + 1}} \left(m - \frac{1}{2} \right) \hbar \left| \ell \frac{1}{2} m + \frac{1}{2} - \frac{1}{2} \right\rangle$$

Now using Eq. (12.6) again and using the orthonormality of states involved, we have

$$\left\langle n\ell + \frac{1}{2}m \,|(L_z + 2S_z)|\, n\ell + \frac{1}{2}m \right\rangle$$

$$= \left(\frac{\ell + m + \frac{1}{2}}{2\ell + 1} \right) \left(m + \frac{1}{2} \right) \hbar$$

$$+ \left(\frac{\ell - m + \frac{1}{2}}{2\ell + 1} \right) \left(m - \frac{1}{2} \right) \hbar$$

$$= m\hbar \left(1 + \frac{1}{2l + 1} \right) \qquad (12.7)$$

and

$$\left\langle nl - \frac{1}{2}m \,|(L_z + 2S_Z)|\, nl - \frac{1}{2}m \right\rangle$$

$$= \left(\frac{\ell - m + \frac{1}{2}}{2\ell + 1} \right) \left(m + \frac{1}{2} \right) \hbar$$

$$+ \left(\frac{\ell + m + \frac{1}{2}}{2\ell + 1} \right) \left(m - \frac{1}{2} \right) \hbar$$

$$= m\hbar \left[1 - \frac{1}{2\ell + 1} \right]$$

$$\therefore \Delta E_{n\ell}^{B} = \langle njm|V|njm \rangle$$

$$= \frac{eB}{2m_e c} \langle njm|(L_z + 2S_z)|njm \rangle$$

$$= \frac{eB}{2m_e c} m\hbar \left[1 \pm \frac{1}{2\ell + 1} \right] \qquad j = \ell \pm \frac{1}{2}$$

Q12.6 If $H(\lambda)$ depends on a parameter λ, and $|\psi_k(\lambda)\rangle$ is an eigenvector of $H(\lambda)$ which is normalized to unity, prove that

$$\frac{dE_k(\lambda)}{d\lambda} = \left\langle \psi_k(\lambda) \left| \frac{\partial H(\lambda)}{\partial \lambda} \right| \psi_k(\lambda) \right\rangle$$

[Helmann-Feynman Theorem]

Solution:

$$\langle \psi_k(\lambda)|H(\lambda)|\psi_k(\lambda)\rangle = E_k(\lambda)\langle \psi_k(\lambda)|\psi_k(\lambda)\rangle$$

$$= E_k(\lambda)$$

$$\frac{dE_k(\lambda)}{d\lambda} = \langle \frac{\partial \psi_k(\lambda)}{\partial \lambda}|H(\lambda)|\psi_k(\lambda)\rangle$$

$$+ \langle \psi_k(\lambda)|\frac{\partial H(\lambda)}{\partial \lambda}|\psi_k(\lambda)\rangle$$

$$+ \langle \psi_k(\lambda)|H(\lambda)|\frac{\partial \psi_k(\lambda)}{\partial \lambda}\rangle$$

$$= E_k(\lambda)\langle \frac{\partial \psi_k(\lambda)}{\partial \lambda}|\psi_k(\lambda)\rangle$$

$$+ E_k(\lambda)\langle \psi_k(\lambda)|\frac{\partial \psi_k(\lambda)}{\partial \lambda}\rangle$$

$$+ \langle \psi_k(\lambda)|\frac{\partial H(\lambda)}{\partial \lambda}|\psi_k(\lambda)\rangle$$

$$= E_k(\lambda)\frac{\partial}{\partial \lambda}\underbrace{\langle \psi_k(\lambda)|\psi_k(\lambda)\rangle}_{=1}$$

$$+ \langle \psi_k(\lambda)|\frac{\partial H(\lambda)}{\partial \lambda}|\psi_k(\lambda)\rangle$$

$$= \langle \psi_k(\lambda)|\frac{\partial H(\lambda)}{\partial \lambda}|\psi_k(\lambda)\rangle$$

Q12.7 Estimate the ground state energy of the hydrogen atom by using the three dimensional harmonic oscillator ground state function as a trial function. How this estimate compares with the exact value, which is $-\frac{me^4}{2\hbar^2}$?

Three dimensional harmonic oscillator ground state function is:

$$\left(\frac{4\alpha^3}{\sqrt{\pi}}\right)^{1/2} e^{-\frac{1}{2}\alpha^2 r^2}, \qquad \alpha = \sqrt{\mu\omega}$$

Take α as variational parameter. Useful integral

$$\int_0^\infty x^{2n} e^{\alpha^2 x^2} dx = \frac{1}{2\alpha^{2n+1}}\frac{\sqrt{\pi}}{n!}\frac{(2n)!}{2^{2n}}.$$

Repeat the same problem with the following trial function:

$$\phi_\alpha(r) = 1 - \frac{r}{\alpha}, \quad r \le \alpha$$

$$= 0, \quad r > \alpha$$

which one is the better trial function?

Solution:

$$E(\alpha) = \frac{\langle \psi(\alpha)|H|\psi(\alpha)\rangle}{\langle \psi(\alpha)|\psi(\alpha)\rangle}$$

$$\langle \psi(\alpha)|\psi(\alpha)\rangle = \frac{4\alpha^3}{\sqrt{\pi}} \int_0^\infty e^{-\alpha^2 r^2}(4\pi r^2)dr$$

$$= 4\pi$$

$$H = -\frac{\hbar^2}{2m}\nabla^2 - \frac{e^2}{r}$$

$$= \frac{\hbar^2}{2m}\left[\frac{1}{r^2}\frac{\partial}{\partial r}\left(r^2\frac{\partial}{\partial r}\right) + \frac{1}{r^2}\Omega(\theta,\phi)\right] - \frac{e^2}{r}$$

$$\Omega(\theta,\phi)[u(r)] = 0$$

Thus

$$E(\alpha) = \frac{\langle H \rangle}{4\pi}$$

$$= \frac{4\alpha^3}{\sqrt{\pi}} \int_0^\infty e^{-\frac{1}{2}\alpha^2 r^2}\left[-\frac{\hbar^2}{2m}\frac{1}{r^2}\frac{\partial}{\partial r}\left(r^2\frac{\partial}{\partial r}\right)\right.$$

$$\left. -\frac{e^2}{r}\right]e^{-\frac{1}{2}\alpha^2 r^2}\cdot 4\pi r^2 dr$$

$$= \frac{4\alpha^3}{\sqrt{\pi}}\left[\left(-3\alpha^2 \int_0^\infty r^2 e^{-\alpha^2 r^2}dr\right.\right.$$

$$\left.+\alpha^4 \int_0^\infty e^{-\alpha^2 r^2}r^4 dr\right)\left(\frac{-\hbar^2}{2m}\right) - e^2 \int_0^\infty e^{-\alpha^2 r^2}r\,dr\right]$$

$$= \frac{3}{2}\alpha^2\left(\frac{-\hbar^2}{2m}\right) - \frac{2\alpha}{\sqrt{\pi}}e^2$$

Now

$$\frac{\partial E(\alpha)}{\partial \alpha} = 3\alpha \left(\frac{\hbar^2}{2m}\right) - \frac{2}{\sqrt{\pi}} e^2 = 0, \quad \text{give}$$

$$\alpha = \frac{4m}{3\sqrt{\pi}\hbar^2} e^2$$

for which

$$\frac{\partial^2 E(\alpha)}{\partial \alpha^2} = \frac{3\hbar^2}{2m} > 0 \tag{12.8}$$

so that the above value of α indicates a maximum point. Hence

$$E_0 = -\frac{8}{3\pi}\left(\frac{me^4}{2\hbar^2}\right)$$

The exact result is

$$E_0 = -\frac{me^4}{2\hbar^2}$$

$$\% \text{ discrepancy } = \frac{\left(1 - \frac{8}{3\pi}\right)}{1} \times 100 = 15\%$$

(b)

$$E(\alpha) = \left[\int_0^a \left(1 - \frac{r}{\alpha}\right)\left(-\frac{\hbar^2}{2m}\right)\right.$$

$$\times \left\{\left[\frac{1}{r^2}\frac{d}{dr}\left(r^2\frac{d}{dr}\right)\right]\left(1 - \frac{r}{\alpha}\right)\right\} r^2 dr$$

$$+ \left.\int_0^a \left(1 - \frac{r}{\alpha}\right)\left(-\frac{e^2}{r}\right)\left(1 - \frac{r}{\alpha}\right) r^2 dr\right]$$

$$\times \frac{1}{\int_0^a \left(1 - \frac{r}{\alpha}\right)^2 r^2 dr}$$

$$= \frac{\left(-\frac{\hbar^2}{2m}\right)\int_0^a \left(r - \frac{r^2}{\alpha}\right) dr + (-e^2)\left(\frac{r^2}{2} - \frac{2}{\alpha}\frac{r^3}{3} + \frac{r^4}{4\alpha^2}\right)\Big|_0^\alpha}{\left(\frac{r^3}{3} - \frac{2}{\alpha}\frac{r^4}{4} + \frac{r^5}{\alpha^2}\right)\Big|_0^\alpha}$$

$$= 5\frac{\hbar^2}{m}\frac{1}{\alpha^2} - \frac{5}{2}e^2\frac{1}{\alpha}$$

$$\frac{\partial E(\alpha)}{\partial \alpha} = -\frac{10\hbar^2}{m\alpha^3} + \frac{5}{2}\frac{e^2}{\alpha^2}$$

$$\frac{\partial E(\alpha)}{\partial \alpha} = 0$$

gives

$$\alpha = \frac{4\hbar^2}{me^2}c$$

For this value of α

$$\frac{\partial^2 E(\alpha)}{\partial \alpha^4} = \frac{30\hbar^2}{m\alpha^2} - 5\frac{e^2}{\alpha^3} > 0$$

Thus the minimum energy is

$$E_0 = -\frac{5}{8}\frac{me^4}{2\hbar^2}$$

$$\%\text{discrepancy} = \frac{\left(1 - \frac{5}{8}\right)}{1} \times 100 \qquad 37.5\%$$

Q12.8 Use the variational principle to estimate the ground state energy of the anharmonic oscillator

$$H = \frac{\hat{p}^2}{2m} + \lambda x^4.$$

Compare your result with the exact result

$$E_0 = 1.060\lambda^{1/3}\left(\frac{\hbar^2}{2m}\right)^{2/3}.$$

You may use the trial function

$$Ae^{-\alpha^2 x^2/2}.$$

Solution:

$$E(\alpha^2) = \frac{\langle\psi|H|\psi\rangle}{\langle\psi|\psi\rangle} \qquad \psi = Ae^{-\frac{\alpha^2 x^2}{2}}$$

$$E(\alpha^2) = \frac{\int_{-\infty}^{+\infty} e^{-\frac{\alpha^2 x^2}{2}}\left(-\frac{\hbar^2}{2m}\frac{\partial^2}{\partial x^2} + \lambda x^4\right)e^{-\frac{\alpha^2 x^2}{2}}\,dx}{\int_{-\infty}^{+\infty} e^{-\alpha^2 x^2}\,dx}$$

$$= \frac{\hbar^2}{2m}\frac{\alpha^2}{2} + \lambda\frac{3}{4}\frac{1}{\alpha^4}$$

$$\frac{dE}{d\alpha^2} = \frac{\hbar^2}{4m} + \frac{3}{4}\lambda(-2)\frac{1}{\alpha^6} = 0$$

gives

$$\alpha^2 = 3^{1/3}\lambda^{1/3}\left(\frac{\hbar^2}{2m}\right)^{-1/3}$$

Then

$$\frac{d^2 E}{d\alpha^4} = -\frac{3\lambda}{2}(-3)\frac{1}{\alpha^8} > 0$$

Thus $E(\alpha)$ has minimum at $\alpha^2 = 3^{1/3}\lambda^{1/3}\left(\frac{\hbar^2}{2m}\right)^{-1/3}$

$$E_0 \approx E_{min} = \frac{3}{4}3^{1/3}\lambda^{1/3}\left(\frac{\hbar^2}{2m}\right)^{2/3}$$

$$\approx 1.082\lambda^{1/3}\left(\frac{\hbar^2}{2m}\right)^{2/3}$$

This differs from the exact result by about 2% only.

Q12.9 Use the variational principle to estimate the ground state energy of a particle in the potential

$$V = \infty \quad \text{for} \quad x < 0$$
$$V = cx \quad \text{for} \quad x > 0$$

Take Axe^{-iax} as the trial function. Why can't one select Ae^{-ax} as the trial function? Useful integral:

$$\int_0^\infty x^n e^{-bx}\,dx = \frac{n!}{b^{n+1}}.$$

Solution: In order to satisfy the boundary condition at $x = 0$, i.e. $\psi = 0$ at $x = 0$ we cannot select Ae^{-ax} as trial wave function since it does not vanish at $x = 0$. Now

$$H = -\frac{\hbar^2}{2m}\frac{d^2}{dx^2} + V$$

Now

$$E(\alpha) = \frac{\langle\psi|H|\psi\rangle}{\langle\psi|\psi\rangle}, \qquad \psi = 0 \text{ for } x < 0$$

$$= \frac{\int_0^\infty xe^{-ax}\left(-\frac{\hbar^2}{2m}\frac{d^2}{dx^2} + cx\right)xe^{-ax}dx}{\int_0^\infty xe^{-ax}xe^{-ax}dx}$$

$$= \frac{\hbar^2}{2m}a^2 + \frac{3}{2}\frac{c}{a}$$

Thus

$$\frac{\partial E(a)}{\partial a} = 0, \qquad \text{gives}$$

$$a = \left(3c\frac{m}{2\hbar^2}\right)^{1/3}$$

Thus

$$E = 3\left(\frac{2\hbar^2c^2}{3m}\right)^{1/3}$$

Chapter 13

Time Dependent Perturbation Theory

Q13.1 A harmonic oscillator is perturbed by a weak, time dependent but spatially uniform force $F(t)$, where $F(t) = 0$ for $t < 0$. Initially, (i.e. $t < 0$) the system is in its ground state $|0\rangle$. Up to first order in F, show that the probability of finding the system in state $|1\rangle$ at time t is given by

$$P(0,1) = \frac{1}{2m\hbar\omega} \left| \int_0^t e^{i\omega t} F(t) dt \right|^2 .$$

For $t \geq 0$, take $F(t) = F_0(1 - e^{-t/\tau})$. Show that in this case, the probability of finding the system in $|1\rangle$ for large times is

$$P(0,1) \approx \frac{F_0^2}{2m\hbar\omega^3},$$

if $\omega\tau \gg 1$. If $\tau = 0$, show that this probability oscillates indefinitely.

Solution:

$$F(t) = -\frac{\partial V}{\partial x}$$

$$V = \begin{cases} -F(t)x & t > 0 \\ 0 & t < 0 \end{cases} \quad \lambda = 1,$$

207

Now

$$\langle 1|V|0 \rangle = V_{10}(t) = -F(t)\langle 1|x|0 \rangle$$

$$= -F(t)\sqrt{\frac{\hbar}{2m\omega}}, \qquad \text{from problem 9.5}$$

Thus

$$P(0,1) = \frac{1}{\hbar^2}\left|\int_0^t e^{i\omega t'}V_{10}(t')dt'\right|^2$$

$$= \frac{1}{2m\hbar\omega}\left|\int_0^t e^{i\omega t'}F(t')dt'\right|^2 \qquad (13.1)$$

Now

$$F(t) = F_0(1 - e^{-t/\tau})$$

and

$$\int_0^t e^{i\omega t'}F_0(1 - e^{-t'/\tau})dt'$$

$$= F_0\left[\frac{e^{i\omega t'}}{i\omega} - \frac{e^{i\omega t' - t'/\tau}}{i\omega - \frac{1}{\tau}}\right]_0^t$$

$$= F_0\left[\frac{e^{i\omega t} - 1}{i\omega} + \tau\frac{e^{i\omega t - t/\tau}}{i\omega\tau - 1} - \frac{\tau}{i\omega\tau - 1}\right] \qquad (13.2)$$

For large t

$$e^{-t/\tau} \to 0$$

Thus for t large and $\omega\tau \gg 1$ (using Eqs. (13.1) and (13.2))

$$P(0,1) \approx \frac{1}{2m\hbar\omega}\left|\frac{F_0 e^{i\omega t}}{i\omega}\right|^2 = \frac{1}{2m\hbar\omega}\frac{F_0^2}{\omega^2}$$

For $\tau = 0$, from Eqs. (13.1) and (13.2)

$$P(0,1) = \frac{1}{2m\hbar\omega}\frac{F_0^2}{\omega^2}|(e^{i\omega t} - 1)|^2$$

$$= \frac{1}{2m\hbar\omega}\frac{4F_0^2}{\omega^2}\sin^2\frac{1}{2}\omega t$$

Q13.2 A hydrogen atom in its ground state is placed in an electric field $\mathbf{E}(t)$, which is spatially uniform and has the time dependence

$$\mathbf{E}(t) = 0 \qquad t < 0$$
$$= \mathbf{E}_0 e^{-t/\tau} \quad t > 0.$$

What is the first-order probability of finding the atom in the $2p$ state after a long time?

Solution:

$$\mathbf{E}(t) = \begin{cases} 0 & t < 0 \\ \mathbf{E}_0 e^{-t/\tau} & t > 0 \end{cases}$$

Take $\mathbf{E}_0$ along z-axis so that

$$E_z = -\frac{\partial V}{\partial z}$$
$$V = -E_0 z e^{-t/\tau} = E(t)z$$
$$\langle 2p|V|1s \rangle = \langle 2p|z|1s \rangle E(t)$$

Thus

$$P(1s \rightarrow 2p) = \left| \frac{i}{\hbar} \int_0^t e^{i\omega t} \langle 2p|z|1s \rangle E(t) dt \right|^2$$

$$= \frac{1}{\hbar^2} E_0^2 |\langle 2p|z|1s \rangle|^2 \left| \int_0^t e^{i\omega t} e^{-t/\tau} dt \right|^2$$

for large t (see previous problem) the value of integral is given below

$$\approx \frac{E_0^2 \tau^2}{(1 + \omega^2 \tau^2)\hbar^2}$$

Now

$$\langle 2p|z|1s \rangle = \int u_{2p} r \cos\theta \, u_{1s} d^3 r$$

$$= \int R_{21} r R_{10} r^2 dr \int Y_{1m}^* \cos\theta \, Y_{00} d\Omega$$

Now (only $m = 0$ gives the nonzero value)

$$\int Y_{10}^* \cos\theta Y_{00} d\Omega = \left(\frac{3}{4\pi}\right)^{1/2} \frac{1}{\sqrt{4\pi}} \int_0^\pi \cos^2\theta 2\pi \sin\theta d\theta$$

$$= \frac{1}{\sqrt{3}} \quad \begin{cases} \text{since } Y_{10} = \sqrt{\frac{3}{4\pi}} \cos\theta \\ Y_{00} = \sqrt{\frac{1}{4\pi}} \end{cases}$$

Thus [see sec. 7.3 of the text]

$$\langle 2p|z|1s\rangle = \frac{1}{\sqrt{3}} \int_0^\infty R_{21} r R_{10} r^2 dr$$

$$= \frac{2}{3}\left(\frac{1}{a}\right)^3 \frac{1}{(2)^{3/2}}\left(\frac{1}{a}\right) \int_0^\infty r^4 e^{-3r/2a} dr$$

Using the formula

$$\int_0^\infty x^n e^{-bx} dx = \frac{n!}{(b)^{n+1}} \quad b = \left(\frac{3}{2a}\right)$$

$$\langle 2p|z|1s\rangle = \frac{2}{3}\frac{4!}{(2)^{3/2}}\frac{2^5}{3^5}a$$

and thus

$$P(1s \to 2p) = \frac{E_0^2 \tau^2}{(1+\omega^2\tau^2)\hbar^2}\frac{4}{9}\frac{(4!)^2}{2^3}\frac{2^{10}}{3^{10}}a^2$$

Additional Problems:

Q1 A hydrogen atom in its ground state is placed in an electric field $\mathbf{E} = \mathbf{E}_o \sin(\omega t)$ of angular frequency $\omega > \frac{me^4}{2\hbar}$. Find the probability per unit time that the atom will be ionized. The functions of the electron in the ionized states may be taken to be plane waves.

Solution: We write the perturbative potential in the form

$$V_t = -e\mathbf{r} \cdot \mathbf{E}_o \sin(\omega t)$$

$$= \frac{e}{2i}\mathbf{r} \cdot \mathbf{E}_o e^{(-i\omega t)} - \frac{e}{2i}\mathbf{r} \cdot \mathbf{E}_o e^{(i\omega t)}$$

Final state

$$|n\rangle = |\mathbf{k}\rangle \Rightarrow u_k(\mathbf{r}) = \frac{1}{(2\pi)^{\frac{2}{3}}} e^{(i\mathbf{k}\cdot\mathbf{r})}$$

$$\mathbf{p} = \hbar\mathbf{k}, \quad \epsilon_k = \frac{p^2}{2m} = \frac{\hbar^2 k^2}{2m}$$

Now initial state

$$|i\rangle = |100\rangle \Rightarrow u_{100}(\mathbf{r}) = (\pi a_o^3)^{-\frac{1}{2}} e^{(-\frac{r}{a_o})}$$

$\epsilon_o = -\frac{me^4}{2\hbar^2}$. The transition probability is

$$P_{ki} = \left| \frac{1}{i\hbar} \int_{-\infty}^{\infty} dt\, e^{(i(\epsilon_k - \epsilon_o)\frac{t}{\hbar})} \langle \mathbf{k}|V_t|100\rangle \right|^2$$

$$= \left| -\frac{e}{2\hbar} \langle \mathbf{k}|\mathbf{r}\cdot\mathbf{E}_o|100\rangle \int_{-\infty}^{\infty} e^{(i(\frac{(\epsilon_k - \epsilon_o)}{\hbar} - \omega)t)} dt \right.$$

$$\left. + \frac{e}{2\hbar} \langle \mathbf{k}|\mathbf{r}\cdot\mathbf{E}_o|100\rangle \int_{-\infty}^{\infty} e^{(i(\frac{(\epsilon_k - \epsilon_o)}{\hbar} + \omega)t)} dt \right|^2$$

The first and second integral gives respectively

$$2\pi\hbar\delta(\epsilon_k - \epsilon_o - \hbar\omega)$$

and

$$2\pi\hbar\delta(\epsilon_k - \epsilon_o + \hbar\omega)$$

The second δ-function does not contribute since $\epsilon_k - \epsilon_o > 0$, $\hbar\omega > 0$ and the argument of the δ-function cannot be zero. Now

$$\Gamma_i = \frac{\sum_{\mathbf{k}} P_{ki}}{t}$$

involves

$$2\pi\hbar\delta(\epsilon_k - \epsilon_o - \hbar\omega) \cdot 2\pi\hbar\delta(\epsilon_k - \epsilon_o - \hbar\omega)$$

$$= (2\pi\hbar)^2 \frac{1}{\hbar} \frac{t}{2\pi} \delta(\epsilon_k - \epsilon_o - \hbar\omega)$$

where we have put $2\pi\hbar\delta(0) = t$, $(t \to \infty)$ [c.f. Eq. (13.25) of the text]

$$\therefore \frac{P_{\mathbf{k}i}}{t} = 2\pi\hbar\delta(\epsilon_{\mathbf{k}} - \epsilon_o - \hbar\omega)\frac{e^2}{4\hbar^2}|\langle\mathbf{k}|\mathbf{r}\cdot\mathbf{E}_o|100\rangle|^2$$

To evaluate

$$I = \langle\mathbf{k}|\mathbf{r}\cdot\mathbf{E}_o|100\rangle\frac{1}{(2\pi)^{\frac{3}{2}}}$$

$$= \frac{1}{(\pi a_0^3)^{\frac{1}{2}}}\int e^{(-i\mathbf{k}\cdot\mathbf{r})}e^{(-\frac{r}{a})}(\mathbf{r}\cdot\mathbf{E}_o)d^3r$$

take $\mathbf{k}$ along z-axis. Then

$$\mathbf{r}\cdot\mathbf{k} = rk\cos\theta$$

$$\mathbf{E}_o = E_o(\sin\beta\cos\gamma, \sin\beta\sin\gamma, \cos\beta)$$

$$\mathbf{r}\cdot\mathbf{E}_o = E_o r(\sin\beta\cos\gamma\sin\theta\cos\phi + \sin\beta\sin\gamma\sin\theta\sin\phi$$
$$+ \cos\beta\cos\theta)$$

$$d^3r = r^2 dr\sin\theta d\theta d\phi$$

$$\int_0^{2\pi}\cos\phi d\phi = 0, \qquad \int_0^{2\pi}\sin\phi d\phi = 0, \qquad \int_0^{2\pi}d\phi = 2\pi$$

Thus we are left with

$$I = \frac{1}{(2a_o)^{\frac{2}{3}}\pi^2}E_o$$

$$\times \int_0^\infty\int_0^\pi e^{-ikr\cos\theta}r\cos\beta\cos\theta e^{-\frac{r}{a}}2\pi\sin\theta d\theta r^2 dr$$

$$= \frac{2\pi}{(2a_o)^{\frac{2}{3}}\pi^2}E_o\cos\beta\left[\int_0^\infty r^3 e^{-\frac{r}{a}}dr\int_{-1}^1 e^{-ikr\eta}\eta d\eta\right]$$

with $\eta = \cos\theta$.
Now

$$\int_{-1}^1 e^{-ikr\eta}\eta d\eta = \left[\frac{1}{-ikr}e^{-ikr\eta}\eta - \left(\frac{1}{-ikr}\right)^2 e^{-ikr\eta}\right]_{-1}^1$$

$$= \frac{1}{ikr}[e^{ikr} - e^{-ikr}] + \left(\frac{1}{ikr}\right)^2[e^{-ikr} + e^{ikr}]$$

∴ we have to evaluate

$$\int_0^\infty r^2 e^{ikr-\frac{r}{a_o}} = r^2 \frac{e^{ikr-\frac{r}{a_o}}}{ik - \frac{1}{a_o}}\Bigg|_0^\infty - 2\int_0^\infty \frac{e^{ikr-\frac{r}{a_o}}}{ik - \frac{1}{a_o}} r\,dr$$

$$= -2\left[\frac{e^{ikr-\frac{r}{a_o}}}{\left(ik - \frac{1}{a_o}\right)^2}r\Bigg|_0^\infty - \frac{ikr - \frac{r}{a_o}}{\left(ik - \frac{1}{a_o}\right)^3}\Bigg|_0^\infty\right]$$

$$= -2\frac{1}{\left(ik - \frac{1}{a_o}\right)^3}$$

$$\int_0^\infty re^{\left(ikr-\frac{r}{a_o}\right)} = \frac{e^{\left(ikr-\frac{r}{a_o}\right)}}{\left(ik - \frac{1}{a_o}\right)}r\Bigg|_0^\infty - \frac{e^{\left(ikr-\frac{r}{a_o}\right)}}{\left(ik - \frac{1}{a_o}\right)^2}\Bigg|_0^\infty$$

$$= \frac{1}{\left(ik - \frac{1}{a_o}\right)^2}$$

and ones obtained by changing $k = -k$.
Thus

$$I = \frac{2\pi}{4\pi^2 a_o^{\frac{3}{2}}} E_o \cos\beta \left[-\frac{1}{ik}\left[\frac{-2}{\left(ik - \frac{1}{a_o}\right)^3} + \frac{-2}{\left(-ik - \frac{1}{a_o}\right)^3}\right]\right.$$

$$\left. + \frac{1}{k^2}\left[\frac{1}{\left(-ik - \frac{1}{a_o}\right)^2} - \frac{1}{\left(ik - \frac{1}{a_o}\right)^2}\right]\right]$$

$$= \frac{2\pi}{4\pi^2 a_o^{\frac{3}{2}}} E_o \cos\beta$$

$$\times \left[\frac{2}{ik} \frac{\left[\left(ik^3 + (-k^2)\left(-\frac{1}{a_o}\right) + 3(-ik)\left(-\frac{1}{a_o}\right)^2 - \frac{1}{a_o^3}\right)\right]}{\left(k^2 + \frac{1}{a_o^2}\right)^3}\right.$$

$$+ \frac{1}{k^2} \frac{\left[\left(-k^2 - 2ik\frac{1}{a_o} + \frac{1}{a_o^2}\right) - (k \to -k)\right]}{\left(k^2 + \frac{1}{a_o^2}\right)^2}$$

$$= \frac{-16iE_o \cos\beta}{\sqrt{2\pi}a_o^{\frac{5}{2}}} \frac{k}{\left(\frac{1}{a_o^2} + k^2\right)^3}$$

Finally thus

$$|I|^2 = \frac{128E_o^2 \cos^2\beta}{\pi^2 a_o^5} \frac{k^2}{\left(\frac{1}{a_o^2} + k^2\right)^6}$$

$$\sum_{\mathbf{k}} = \int d^3k = \int k^2 dk d\Omega_k$$

$$d\Gamma_i = \frac{\Sigma_k P_{\mathbf{k}} i}{t}$$

$$= \frac{128E_o^2 \cos^2\beta}{\pi^2 a_o^2} d\Omega_k$$

$$\times \int 2\pi\hbar \frac{e^2}{4\hbar^2} \delta(\epsilon_{\mathbf{k}} - \epsilon_o - \hbar\omega) \frac{k^3}{\left(\frac{1}{a_o^2} + k^2\right)^6} \frac{m}{\hbar^2} d\epsilon_k$$

$$= \frac{64e^2 E_o^2 \cos^2\beta}{\pi^2 a_o^5} d\Omega_k \frac{mk^3}{\hbar^3} \frac{k^3}{\left(\frac{1}{a_o^2} + k^2\right)^6}$$

where it is now understood that

$$k = \sqrt{\frac{2m}{\hbar}(\omega - \omega_o)}$$

Now

$$a_o^2 = \frac{\hbar^4}{m^2 e^4}, \quad k^2 a_o^2 = \frac{2m}{\hbar}(\omega - \omega_o)\frac{\hbar^4}{m^2 e^4} = \frac{(\omega - \omega_o)}{m\frac{e^4}{2\hbar^3}}$$

$$= \frac{(\omega - \omega_o)}{\omega} = \left(\frac{\omega}{\omega_o} - 1\right)$$

Thus

$$dΓ_i = \frac{64e^2 E_o^2 \cos^2 β}{π^2 a_o^5} \frac{m(\frac{2m}{ℏ}ω)^{\frac{3}{2}}(\frac{ω}{ω}-1)^{\frac{3}{2}} a_o^{12}}{ℏ^3(\frac{ω}{ω_o}+1)^6} dΩ_k$$

$$= \frac{64a_o^3 E_o^2}{πℏ} \left(\frac{ω_o}{ω}\right)^6 \left(\frac{ω}{ω_o}-1\right)^{\frac{3}{2}} \cos^2 β dΩ_k$$

where we have used $a_o^2 = \frac{ℏ^4}{m^2 e^4}$.

Now

$$dΩ_k = 2π \sin β dβ$$

$$\int_0^π \cos^2 β \sin β dβ = -\frac{\cos^3 β}{3}\bigg|_0^π = \frac{2}{3}$$

Thus finally

$$Γ = \frac{256 a_o^3 E_o^2}{3ℏ} \left(\frac{ω_o}{ω}\right)^6 \left(\frac{ω}{ω_o}-1\right)^{\frac{3}{2}}$$

Chapter 14

Statistics and the Exclusion Principle

Q14.1 A deuteron consists of a neutron and a proton. Consider a system of 2 deuterons as a composite of 2 neutrons and 2 protons with the wave function

$$\psi(n_2, p_2; n_1, p_1)$$

where n and p denote *all* the coordinates of a neutron and a proton respectively. How does the above wave function behave under the interchange of 2 deuterons, i.e. under

$$(n_1, p_1) \leftrightarrow (n_2, p_2)?$$

Solution:

$$\psi(n_2, p_2; n_1, p_1) = -\psi(n_1, p_2; n_2, p_1)$$
$$= \psi(n_1, p_1; n_2, p_2)$$

Deuteron wave function is symmetric; it is a boson

Q14.2 For a system of 2 identical particles, is

$$\psi(1, 2) = \psi_{100}(r_1)\psi_{200}(r_2)$$

where

$$\psi_{100}(r) = \frac{1}{\sqrt{2\pi}}\beta^{3/2}e^{-\beta r},$$

$$\psi_{200}(r) = \frac{1}{4\sqrt{2\pi}}\beta^{3/2}(2 - \beta r)e^{-\beta r},$$

an acceptable wave function? If not, find an acceptable wave function. If the particles have spin $\frac{1}{2}$, write down acceptable wave functions which include the spin wave functions also.

Solution:

$$\psi(1,2) = \psi_{100}(r_1)\psi_{200}(r_2)$$

is not an acceptable wave function.
Symmetric and antisymmetric wave functions are

$$\psi_{s,a} = \frac{1}{2}[\psi(1,2) \pm \psi(2,1)]$$

$$= \frac{1}{2}[\psi_{100}(r_1)\psi_{200}(r_2) \pm \psi_{100}(r_2)\psi_{200}(r_1)]$$

$$= \frac{1}{8\pi}\beta^3[e^{-\beta r_1}(2 - \beta r_2)e^{-\beta r_2}$$

$$\pm e^{-\beta r_2}(2 - \beta r_1)e^{-\beta r_1}]$$

$$\psi_s(1,2) = \frac{1}{8\pi}\beta^3 e^{-\beta(r_1+r_2)}[2 - \beta(r_1 + r_2)]$$

$$\psi_a(1,2) = \frac{1}{8\pi}\beta^4 e^{-\beta(r_1+r_2)}(r_1 - r_2)$$

Thus for spinless particles, the acceptable wave function is $\psi_s(1,2)$. For spin $\frac{1}{2}$ particles, the acceptable wave function, which has to be antisymmetric, is [χ_- is antisymmetric with respect to spin coordinates while χ_+ is symmetric]

$$\text{either} \quad \frac{1}{8\pi}\beta^3 e^{-\beta(r_1+r_2)}[2 - \beta(r_1 + r_2)]\chi_-(\sigma_1, \sigma_2)$$

$$\text{or} \quad \frac{1}{8\pi}\beta^4 e^{-\beta(r_1+r_2)}(r_1 - r_2) \begin{cases} \chi_+^{+1}(\sigma_1, \sigma_2) \\ \chi_+^{0}(\sigma_1, \sigma_2) \\ \chi_+^{-1}(\sigma_1, \sigma_2) \end{cases}$$

Q14.3 For a helium atom $(Z = 2)$, show that the ground state energy to first order in perturbation is

$$E_0 = E_0^{(0)} + \Delta E$$

where

$$E_0^{(0)} = -\frac{Z^2 e^2}{a_0}, \quad \Delta E = (5/8)\frac{Ze^2}{a_0}$$

Hint: Ground state wave function is

$$\psi(1,2) = u_{100}(\mathbf{r}_1)u_{100}(\mathbf{r}_2)\chi_-(\sigma_1,\sigma_2)$$

where

$$u_{100}(\mathbf{r}) = \frac{1}{\sqrt{\pi}}(\frac{Z}{a_0})^{3/2}e^{-Zr/a_0}$$

Apply Eq. (14.33a) of the text with the above wave function.

Solution: For this case

$$\Delta E = \lambda V_{11}$$

$$= \frac{e^2}{\pi^2}\left(\frac{Z}{a_0}\right)^6 \int\int e^{-2Z/a_0(r_1+r_2)}\frac{1}{|\mathbf{r}_1-\mathbf{r}_2|}d^3r_1 d^3r_2$$

But using the Fourier transformation

$$\frac{1}{|\mathbf{r}_1-\mathbf{r}_2|} = \int \frac{d^3k}{(2\pi)^3}e^{i\mathbf{k}\cdot(\mathbf{r}_1-\mathbf{r}_2)}\frac{4\pi}{k^2}$$

$$= \frac{4\pi}{\pi^2}\left(\frac{Z}{a_0}\right)^6 e^2\frac{1}{(2\pi)^3}$$

$$\times \int \frac{d^3k}{k^2}\left(\left(d^3r_1 e^{i\mathbf{k}\cdot\mathbf{r}_1}e^{-\frac{2Z}{a_0}r_1}\right)\right.$$

$$\times \left.\int d^3r_2 e^{-i\mathbf{k}\cdot\mathbf{r}_2}e^{-\frac{2Z}{a_0}r_2}\right)$$

$$= \frac{1}{(2\pi)^4}\left(\frac{Z}{a_0}\right)^6 e^2 \int \frac{d^3k}{k^2}\left|\int d^3r\, e^{i\mathbf{k}\cdot\mathbf{r}-\frac{2Z}{a_0}r}\right|^2$$

But

$$\int d^3r\, e^{i\mathbf{k}\cdot\mathbf{r}-\frac{2Z}{a_0}r} = \frac{16\pi Z/a_0}{\left[k^2+\left(\frac{2Z}{a_0}\right)^2\right]^2}$$

This can be easily seen by using spherical polar coordinates: $d^3r = 2\pi \sin\theta r^2 dr$, $\mathbf{k} \cdot \mathbf{r} = k\cos\theta$ and then integrating by parts for r-integration and noting that $e^{-\frac{2Z}{a_0}r} \to 0$ as $r \to \infty$ so that the $r = 0$ limit of integral will be relevant. Thus

$$\Delta E = \frac{1}{2\pi^4}\left(\frac{Z}{a_0}\right)^6 e^2(16\pi)^2 \left(\frac{Z}{a_0}\right)^2$$

$$\times \int_0^\infty \frac{4\pi k^2 dk}{\left[k^2 + \left(\frac{2Z}{a_0}\right)^2\right]^4} \frac{1}{k^2}$$

By change of variable $\frac{a_0}{2Z}k = x$, we can write finally

$$\Delta E = \frac{4}{\pi}\frac{Ze^2}{a_0} \int_0^\infty \frac{dx}{(x^2 + 1)^4}$$

Again by change of variable $x = \tan\theta$, we have

$$\Delta E = \frac{4}{\pi}\frac{Ze^2}{a_0} \int_0^{\pi/2} \cos^6\theta d\theta$$

$$= \frac{4}{\pi}\frac{Ze^2}{a_0}\frac{5\pi}{32}$$

Q14.4 Consider an electron confined in a cubical box of length L. Show that the energy eigenvalues and normalised eigenfunctions are given by

$$E = \frac{\hbar^2\pi^2}{2mL^2}(n_1^2 + n_2^2 + n_3^2)$$

$$u_n(\mathbf{r}) = \left(\frac{2}{L}\right)^{2/3} \sin\frac{n_1\pi}{L}x \sin\frac{n_2\pi}{L}y \sin\frac{n_3\pi}{L}z.$$

From this it follows that the number of quantum states between momentum $\mathbf{p}$ and $\mathbf{p} + d\mathbf{p}$ is given by [$\mathbf{p} = \hbar\mathbf{k}$ and

$\frac{2\pi n_1}{L} = k_x$ etc.]

$$2\left(\frac{L}{2\pi}\right)^3 dk_x dk_y dk_z = 2\left(\frac{L}{2\pi}\right)^3 d^3k$$

$$= 2\left(\frac{L}{2\pi}\right)^3 4\pi k^2 dk,$$

where the factor 2 is due to two spin orientation. Consider now a non-interacting electron gas in the box. A quantum state is specified by three quantum numbers (n_1, n_2, n_3). Since an electron has spin 1/2, only two electrons can be in the same state specified by (n_1, n_2, n_3). Now the Pauli principle forces the electrons to occupy all the states between 0 and ϵ_F, where ϵ_F is the Fermi energy and for non-relativistic case is given by $\epsilon_F = \frac{P_F^2}{2m}$, show that

$$\epsilon_F = \frac{\pi^2 \hbar^2}{2m}\left(\frac{3\rho}{m\pi}\right)^{2/3}$$

where $\rho = \frac{N}{L^3}m$, is the density of electrons in the box. Find the pressure as a function of ρ viz. the equation of state for the degenerate gas. Obtain also the equation of state for a relativistic degenerate electron gas ($\epsilon_F \approx cp_F$).

Solution: The Schrödinger equation inside the box is [c.f. Eq. 4.52 of the text]

$$-\frac{\hbar^2}{2m}\nabla^2 u = Eu$$

$$-\frac{\hbar^2}{2m}\left[\frac{\partial^2}{\partial x^2} + \frac{\partial^2}{\partial y^2} + \frac{\partial^2}{\partial z^2}\right]u = Eu$$

This is separable and we can write

$$u(x, y, z) = u_1(x)u_2(y)u_3(z)$$

where

$$-\frac{\hbar^2}{2m}\frac{\partial^2}{\partial x^2}u_1(x) = E_1 u_1(x)$$

$$-\frac{\hbar^2}{2m}\frac{\partial^2}{\partial y^2}u_2(y) = E_2 u_2(y)$$

$$-\frac{\hbar^2}{2m}\frac{\partial^2}{\partial z^2}u_3(z) = E_3 u_3(y)$$

where $E_1 + E_2 + E_3 = E$. Thus using Eq. (4.55) $(2a = L)$,

$$u_n(\mathbf{r}) = \left(\frac{2}{L}\right)^3 \sin\frac{n_1\pi x}{L}\sin\frac{n_2\pi y}{L}\sin\frac{n_3\pi z}{L}$$

with

$$E = \frac{\hbar^2}{2mL^2}(n_1^2 + n_2^2 + n_3^2) \qquad (14.1)$$

$$dn = 2\left(\frac{L}{2\pi\hbar}\right)^3 4\pi p^2 dp$$

$$n_1^2 + n_2^2 + n_3^2 = \frac{2mL^2 E}{\hbar^2\pi^2}$$

For fixed E. This is an equation of a sphere of radius $R = \sqrt{\frac{2mL^2}{\hbar^2\pi^2}E}$

Hence total number of states

$$N = 8\pi\left(\frac{L}{2\pi\hbar}\right)^3 \int_0^{p_F} p^2 dp$$

$$= 8\pi\left(\frac{L}{2\pi\hbar}\right)^3 \frac{p_F^3}{3}$$

Each state is occupied by one electron. N is number of electrons.

$$\epsilon_F = \frac{p_F^2}{2m}$$

$$N = 8\pi\left(\frac{L}{2\pi\hbar}\right)^3 \frac{(2m\epsilon_F)^{3/2}}{3}$$

$$\epsilon_F = \left(\frac{3N}{8\pi V}\right)^{2/3}\frac{(2\pi\hbar)^2}{2m}$$

Now

$$\rho = \frac{N}{L^3}m = \frac{N}{V}m$$

Thus

$$\epsilon_F = \frac{\pi^2\hbar^2}{2m}\left(\frac{3\rho}{\pi m}\right)^{2/3}$$

Now

$$E = \int_0^{p_F} \frac{p^2}{2m}dn$$

$$= \frac{1}{2m}2\frac{V}{(2\pi\hbar)^3}4\pi\int_0^{p_F}p^4dp$$

$$= \frac{V4\pi}{m(2\pi\hbar)^3}\frac{p_F^5}{5} = \frac{4\pi V}{m(2\pi\hbar)^3}\frac{(2m\epsilon_F)^{5/2}}{5}$$

$$= \frac{4\pi V}{5m}(2\pi\hbar)^2\left(\frac{3N}{8\pi V}\right)^{5/3}$$

$$= \frac{6\pi^2\hbar^2}{5m}\left(\frac{3}{8\pi}\frac{N}{V}\right)^{2/3}N = \frac{3}{5}N\epsilon_f$$

So

$$P = -\frac{\partial E}{\partial V}$$

$$= \frac{4\pi^2\hbar^2}{5m}\left(\frac{3}{8\pi}\right)^{2/3}\left(\frac{\rho}{m}\right)^{5/3}$$

$$= \frac{\hbar^2}{5m}(3\pi^2)^{2/3}\left(\frac{\rho}{m}\right)^{5/3}$$

Now for relativistic gas, using $\epsilon_F \approx cp_F$

$$N = 8\pi\left(\frac{L}{2\pi\hbar}\right)^3\frac{\epsilon_F^3}{3c^3}$$

$$= \frac{1}{3\pi^2}\frac{V}{(c\hbar)^3}\epsilon_F^3$$

Thus

$$\epsilon_F = (3\pi^2)^{1/3}(c\hbar)\left(\frac{N}{V}\right)^{1/3}$$

Thus for relativistic gas

$$E = \int_0^{\epsilon_F} \epsilon \, dn$$

$$= 2V \frac{4\pi}{(2\pi c\hbar)^3} \int_0^{\epsilon_F} \epsilon^3 d\epsilon$$

$$= V \frac{2\pi}{(2\pi c\hbar)^3} \epsilon_F^4 \tag{14.2}$$

where we have used

$$\epsilon = cp$$

$$dp = \frac{1}{c} d\epsilon$$

Hence

$$E = V \frac{c\hbar}{4\pi^2}(3\pi^2)^{4/3}\left(\frac{N}{V}\right)^{4/3}$$

$$= \frac{3}{4} N \epsilon_F$$

$$\therefore P = -\frac{\partial E}{\partial V} = \frac{1}{3}\frac{E}{V} = \frac{1}{3}\rho_E = \frac{1}{3}\rho c^2$$

Q14.5 Consider two electrons confined in a box whose sides are of length L. There exists an attractive potential of strength V_0 between pairs of electrons whenever they are very close to each other. Using perturbation theory, calculate the ground state energy and the wave function. Hint: Take the potential

$$V(\mathbf{r}_1 - \mathbf{r}_2) = -V_0 \qquad r < a$$
$$= 0 \qquad r > a, a \ll L.$$

Then approximate this potential as

$$\int V(\mathbf{r}_1 - \mathbf{r}_2)d^3r_2 = -V_0\frac{4\pi}{3}a^3$$

$$V(\mathbf{r}_1 - \mathbf{r}_2) = -V_0\frac{4\pi}{3}a^3\delta(\mathbf{r}_1 - \mathbf{r}_2).$$

Treat this potential as a perturbation. Then using equations similar to Eqs. (14.33), (14.34), (14.36), (14.37), (14.39) and (14.41) in the text and the result of first part of problem (14.4), calculate energy levels and energy eigenfunctions. For a numerical estimate take $L = 10^{-8}$ cm, $a < 10^{-10}$ cm and $V_0 = 10^{-3}$ eV.

Solution: For a particle in a box of length L, the energy eigenvalue and eigenfunctions are given by

$$E_\alpha = \frac{\pi^2 \hbar^2}{2mL^2}(n_{\alpha1}^2 + n_{\alpha2}^2 + n_{\alpha3}^2)$$

$$u_\alpha(\mathbf{r}) = \left(\frac{2}{L}\right)^{3/2} \sin \frac{n_{\alpha1}\pi}{L} x \sin \frac{n_{\alpha2}\pi}{L} y \sin \frac{n_{\alpha3}\pi}{L} z$$

For a two-electron system, the total wave function should be antisymmetric under the exchange of electrons. Thus the allowed wave functions are

$$\frac{1}{\sqrt{2}}\left[u_\alpha(\mathbf{r}_1 u_\beta(\mathbf{r}_2) + u_\beta(\mathbf{r}_1)u_\alpha(\mathbf{r}_2)\right]\chi_-(1,2) : s = 0$$

$$(14.3)$$

$$\frac{1}{\sqrt{2}}\left[u_\alpha(\mathbf{r}_1)u_\beta(\mathbf{r}_2) - u_\beta(\mathbf{r}_1)u_\alpha(\mathbf{r}_2)\right]\chi_+(1,2) : s = 1$$

$$(14.4)$$

where $\chi_-(1,2)$ and $\chi_+(1,2)$ are antisymmetric and symmetric wave function respectively.
For the perturbation

$$V(\mathbf{r}_1 - \mathbf{r}_2) = -V_0 \frac{4\pi}{3} a^3 \delta(\mathbf{r}_1 - \mathbf{r}_2)$$

the second wave function would vanish at $\mathbf{r}_1 = \mathbf{r}_2$, as the perturbative potential contribute for $\mathbf{r}_1 = \mathbf{r}_2$. This means for the symmetric spin state for $s = 1$, the contribution of perturbation vanishes and there is no shift for this energy levels. For spin-singlet state, the shift is energy is given by

$$\Delta E = \int u_+^* V(\mathbf{r}_1 - \mathbf{r}_2)u_+ d^3 r_1 d^3 r_2$$

$$= -V_0 \frac{4\pi}{3} \int (u_+(\mathbf{r}_1))^2 d^3 r_1$$

$$= V_0 \frac{4\pi}{3} a^3 \int [u_\alpha(\mathbf{r}_1)]^2 [u_\beta(\mathbf{r}_1)]^2 d^3 r_1$$

In order to evaluate the integral of the form

$$\int_0^L \left[\sin \frac{n_{\alpha 1} \pi x}{L} \sin \frac{n_{\beta 1} \pi x}{L} \right]^2 dx$$

$$= \frac{L}{\pi} \int_0^L (\sin^2 n_{\alpha 1} \theta \sin^2 n_{\beta 1} \theta) d\theta$$

$$= \left(\frac{L}{4\pi} \right) \frac{1}{2} \int_0^{2\pi}$$

$$\times (\cos n_{\alpha 1} \phi - 1)(\cos n_{\beta 1} \phi - 1) \text{ where } \phi = 2\theta$$

$$= \left(\frac{L}{4\pi} \right) \frac{1}{2} (2\pi)$$

$$= \frac{L}{4} \quad \text{since the cosine terms give zero}$$

Hence

$$\Delta E = \left[-V_0 \frac{4\pi}{3} a^3 \right]$$

$$= \left(\frac{2}{L} \right)^6 \left(\frac{L}{4} \right)^3 \left[V_0 \frac{4\pi}{3} a^3 \right]$$

$$= -V_0 \frac{4\pi}{3} \frac{a^3}{L^3}$$

Thus for $s = 0$, i.e. for spin singlet state

$$E = E_\alpha + E_\beta$$

$$= \frac{\pi^2 \hbar^2}{2mL^2} [(n_{\alpha 1}^2 + n_{\alpha 2}^2 + n_{\alpha 3}^2)$$

$$+ (n_{\beta 1}^2 + n_{\beta 2}^2 + n_{\beta 3}^2)]$$

$$+ 2 \left(-V_0 \frac{4\pi}{3} \frac{a^3}{L^3} \right)$$

Hence for the ground state

$$E = \frac{3\pi^2 \hbar^2}{mL^2} - \frac{8\pi V_0}{3} \frac{a^3}{L^3}$$

Chapter 15

Two State Systems

Q15.1 Show that $\begin{pmatrix} \cos\frac{\theta}{2} \\ e^{i\phi}\sin\frac{\theta}{2} \end{pmatrix}$ and $\begin{pmatrix} -e^{i\phi}\sin\frac{\theta}{2} \\ \cos\frac{\theta}{2} \end{pmatrix}$ are eigenstates of the helicity operator $\frac{\sigma\cdot\mathbf{n}}{|\mathbf{n}|}$ with eigenvalues ± 1 respectively. Here $\frac{\mathbf{n}}{|\mathbf{n}|} = (\sin\theta\cos\phi, \sin\theta\sin\phi, \cos\theta)$.

Solution:

$$\mathbf{n}\cdot\boldsymbol{\sigma} = \sin\theta\cos\phi\,\sigma_x + \sin\theta\sin\phi\,\sigma_y + \cos\theta\,\sigma_z$$

$$= \begin{pmatrix} \cos\theta & \sin\theta e^{-i\phi} \\ \sin\theta e^{i\phi} & -\cos\theta \end{pmatrix}$$

Thus

$$\mathbf{n}\cdot\boldsymbol{\sigma}\begin{pmatrix} \cos\frac{\theta}{2} \\ e^{i\phi}\sin\frac{\theta}{2} \end{pmatrix} = \begin{pmatrix} \cos\theta\cos\frac{\theta}{2} + \sin\frac{\theta}{2}\sin\theta \\ e^{i\phi}(\sin\theta\cos\frac{\theta}{2} - \cos\theta\sin\frac{\theta}{2}) \end{pmatrix}$$

Using $\cos\theta = \cos^2\frac{\theta}{2} - \sin^2\frac{\theta}{2} = 1 - 2\sin^2\frac{\theta}{2}$, $\sin\theta = 2\sin\frac{\theta}{2}\cos\frac{\theta}{2}$, at the appropriate places,

$$\mathbf{n}\cdot\boldsymbol{\sigma}\begin{pmatrix} \cos\frac{\theta}{2} \\ e^{i\phi}\sin\frac{\theta}{2} \end{pmatrix} = \begin{pmatrix} \cos\frac{\theta}{2} \\ e^{i\phi}\sin\frac{\theta}{2} \end{pmatrix}$$

Similarly

$$\mathbf{n}\cdot\boldsymbol{\sigma}\begin{pmatrix} -e^{-i\phi}\sin\frac{\theta}{2} \\ \cos\frac{\theta}{2} \end{pmatrix} = -\begin{pmatrix} -e^{i\phi}\sin\frac{\theta}{2} \\ \cos\frac{\theta}{2} \end{pmatrix}$$

Q15.2 Show that the probability of finding the particle in eigenstates of σ_x with eigenvalues ± 1 is

$$P_{x\uparrow\downarrow} = \frac{1}{2}\left\{\frac{1}{2}1 \pm 2\left(\sin^2\frac{\Omega}{2}t\sin\theta\cos\theta\cos\omega t\right.\right.$$
$$\left.\left. + \sin\frac{\Omega}{2}t\cos\frac{\Omega}{2}t\sin\theta\sin\omega t\right)\right\}.$$

What is the corresponding result for $P_{y\uparrow\downarrow}$.
Hint: Eigenstates of σ_x and σ_y are respectively

$$\frac{1}{\sqrt{2}}\left[\left|+\frac{1}{2}\right\rangle \pm \left|-\frac{1}{2}\right\rangle\right], \quad \frac{1}{\sqrt{2}}\left[\left|+\frac{1}{2}\right\rangle \pm i\left|-\frac{1}{2}\right\rangle\right]$$

Solution: Using the Eq. (15.47) of the text, the probability of finding the particle in eigenstate of σ_x with eigenvalues ± 1 i.e. for the states

$$\frac{1}{\sqrt{2}}\left[\left|+\frac{1}{2}\right\rangle \pm \left|-\frac{1}{\sqrt{2}}\right\rangle\right]$$

is

$$P_{x\uparrow\downarrow} = \left|\frac{1}{\sqrt{2}}\left(\left\langle +\frac{1}{2}\left|\chi(t)\right\rangle \pm \left\langle -\frac{1}{2}\left|\chi(t)\right\rangle\right)\right|^2\right.$$

$$= \frac{1}{2}\left|e^{-i\omega t/2}\left(\cos\frac{\Omega}{2}t - i\sin\frac{\Omega}{2}t\cos\theta\right)\right.$$

$$\left.\mp ie^{i\omega t/2}\sin\frac{\Omega}{2}t\sin\theta\right|^2$$

$$= \frac{1}{2}\left|\cos\frac{\Omega}{2}t - i\sin\frac{\Omega}{2}t\cos\theta\right|^2 + \left|\sin\frac{\Omega}{2}t\sin\theta\right|^2 \quad (15.1)$$

$$\mp 2Re\left[\left(e^{-i\omega t}(-i)\sin\frac{\Omega}{2}t\sin\theta\right)\right.$$

$$\left.\times \left(\cos\frac{\Omega}{2}t - i\sin\frac{\Omega}{2}t\cos\theta\right)\right]$$

$$= \frac{1}{2} \left[1 \mp 2 \sin \frac{\Omega}{2} t \sin \theta \right.$$

$$\left. \times \left(- \cos \omega t \sin \frac{\Omega}{2} t \cos \theta - \sin \omega t \cos \frac{\Omega}{2} t \right) \right]$$

which gives the required result.
Similarly

$$P_{y \uparrow \downarrow} = \frac{1}{2} \left| \left\langle + \frac{1}{2} \middle| \chi(t) \right\rangle \pm i \left\langle - \frac{1}{2} \middle| \chi(t) \right\rangle \right|^2$$

$$= \frac{1}{2} \left[1 \pm \left(\sin \Omega t \sin \theta \cos \omega t \right. \right.$$

$$\left. \left. - 2 \sin^2 \frac{\Omega}{2} t \sin \theta \cos \theta \sin \omega t \right) \right]$$

Chapter 16

Quantum Computation

Q16.1 Show that the CNOT quantum logic gate is a reversible two -bit gate.

Solution:

$$|q_1\rangle \otimes |q_2\rangle \xrightarrow{\hat{U}_{CNOT}} |q_1\rangle \otimes |q_1 \oplus q_2\rangle$$

$$|q_1\rangle \otimes |q_1 \oplus q_2\rangle \xrightarrow{\hat{U}_{CNOT}} |q_1\rangle \otimes |q_1 \oplus q_1 \oplus q_2\rangle$$

$$= |q_1\rangle \otimes |q_2\rangle$$

Hence $\hat{U}_{CNOT}$ is a reversible two qubit logic gate.

Q16.2 Show that Bell states, given as

$$|\alpha_\pm\rangle = \frac{|0,0\rangle \pm |1,1\rangle}{\sqrt{2}},$$

$$|\beta_\pm\rangle = \frac{|0,1\rangle \pm |1,0\rangle}{\sqrt{2}},$$

define a complete set of the basic vectors for a two qubit system, that is, the four entangled states are orthogonal and normalized to unity.

Solution: We calculate $\langle \alpha_j | \beta_k \rangle$ where $j = +$ and $-$, similarly $k = +$ and $-$.

If $j = k$

$$\langle \alpha_j | \beta_j \rangle = 1$$

However for $j \neq k$

$$\langle \alpha_j | \beta_k \rangle = 0$$

Hence $|\alpha_\pm\rangle$ and $|\beta_\pm\rangle$ define a complete set of the basis vectors.

Q16.3 Show that the tensor product of two unit operators a is a unit operator.

Solution:

$$I \otimes I =?$$

where $\hat{I}$ is a unit operator.

If

$$I = \begin{pmatrix} 1 & 0 \\ 0 & 1 \end{pmatrix}$$

then

$$I \otimes I = \begin{pmatrix} I & O \\ O & I \end{pmatrix} = \begin{pmatrix} 1 & 0 & 0 & 0 \\ 0 & 1 & 0 & 0 \\ 0 & 0 & 1 & 0 \\ 0 & 0 & 0 & 1 \end{pmatrix}$$

is a unit operator.

Q16.4 Show that the tensor product of two unitary operators is a unitary operator.

Solution:

$$\hat{U} = \begin{pmatrix} e^{i\theta} & 0 & \cdots \\ 0 & e^{i\theta} & \cdots \\ \vdots & \vdots & \ddots \end{pmatrix}_{n \times n}$$

$$U \otimes U = U \otimes \begin{pmatrix} e^{i\theta} & 0 & \cdots \\ 0 & e^{i\theta} & \cdots \\ \vdots & \vdots & \ddots \end{pmatrix}$$

$$= \begin{pmatrix} e^{2i\theta} & 0 & \cdots \\ 0 & e^{2i\theta} & \cdots \\ \vdots & \vdots & \ddots \end{pmatrix}$$

Chapter 17

Perturbation Induced by
Electromagnetic Field

Q17.1 Let $\mathbf{B}$ be a uniform magnetic field. Verify that $\mathbf{A} = \frac{1}{2}\mathbf{B} \times \mathbf{r}$ satisfies the relation $\mathbf{B} = \nabla \times \mathbf{A}$ and the gauge condition $\nabla \cdot \mathbf{A} = 0$. Take $\mathbf{B} \equiv (0, 0, B)$ and show that $\mathbf{A} \equiv \frac{1}{2}(-By, Bx, 0)$.

Solution:

$$\mathbf{B} = \nabla \times \mathbf{A}$$

$$= \frac{1}{2}\nabla \times (\mathbf{B} \times \mathbf{r})$$

Now

(i)

$$B_i = \frac{1}{2}\epsilon_{ijk}\frac{\partial}{\partial x_j}(\mathbf{B} \times \mathbf{r})_k$$

$$= \frac{1}{2}\epsilon_{ijk}\epsilon_{kj'i'}\frac{\partial}{\partial x_j}(B_{j'}x_{i'})$$

$$= \frac{1}{2}(\delta_{ij'}\delta_{ji'} - \delta_{ii'}\delta_{jj'})B_{j'}\frac{\partial x_{i'}}{\partial x_j}$$

$$= \frac{1}{2}(\delta_{ij'}\delta_{ji'} - \delta_{ii'}\delta_{jj'})B_{j'}\delta_{i'j}$$

$$= \frac{1}{2}(2\delta_{ij'})B_{j'} = B_i$$

235

(ii)

$$\mathbf{A} = \frac{1}{2}\mathbf{B} \times \mathbf{r} = \frac{B}{2}[\mathbf{e}_z \times (x\mathbf{e}_x + y\mathbf{e}_y + z\mathbf{e}_z)]$$

$$= \frac{B}{2}[x\mathbf{e}_y - y\mathbf{e}_x]$$

$$\mathbf{A} = \frac{1}{2}(-By, Bx, 0)$$

Q17.2 Consider the motion of a particle of charge e and mass m in $x - y$ plane with a uniform magnetic field along $z-$axis, $\mathbf{B} = \nabla \times \mathbf{A}$, $\mathbf{A} = \frac{1}{2}\mathbf{B} \times \mathbf{r}$.
Show that

$$\hat{Q} = \frac{1}{m\omega}\left(\hat{p}_x - \frac{e}{c}A_x\right) \tag{17.1}$$

and

$$\hat{P} = \left(\hat{p}_y - \frac{e}{c}A_y\right) \tag{17.2}$$

satisfy

$$[\hat{Q}, \hat{P}] = i\hbar. \tag{17.3}$$

Show that Hamiltonian

$$H = \frac{1}{2m}\left(\hat{\mathbf{p}} - \frac{e}{c}\mathbf{A}\right)^2 \tag{17.4}$$

can be written as

$$H = \frac{\hat{P}^2}{2m} + \frac{1}{2}m\omega^2\hat{Q}^2 \tag{17.5}$$

and hence the energy eigenvalues, called Landau levels are given by

$$E = \left(n + \frac{1}{2}\right)\hbar\omega. \tag{17.6}$$

Show that the other canonical pair

$$\hat{Q}' = \left(\hat{p}_y + \frac{e}{c}A_y\right)$$

$$\hat{P}' = \frac{1}{m\omega}\left(\hat{p}_x + \frac{e}{c}A_x\right) \tag{17.7}$$

are cyclic i.e. they commute with $\hat{Q}$, $\hat{P}$ and do not occur in the Hamiltonian.

Solution: Now

$$[\hat{Q}, \hat{P}] = \frac{1}{m\omega}(-e/c)\left([\hat{p}_x, A_y] - [\hat{p}_y, A_x]\right)$$

$$= \frac{1}{m\omega}(-e/c)(-i\hbar)\left[\frac{\partial A_y}{\partial x} - \frac{\partial A_x}{\partial y}\right]$$

$$= \frac{ie\hbar}{m\omega c}(\nabla \times \mathbf{A})_z = \frac{ie\hbar}{m\omega c}B = i\hbar \qquad (17.8)$$

for $\omega = \frac{eB}{mc}$, they are canonical variables.

Now

$$H = \frac{1}{2m}\left(\hat{\mathbf{p}} - \frac{e}{c}\mathbf{A}\right)^2$$

$$= \frac{\hat{P}^2}{2m} + \frac{1}{2}m\omega^2\hat{Q}^2$$

which is the Hamiltonian of one dimensional harmonic oscillator and has energy eigenvalues

$$E = \left(n + \frac{1}{2}\right)\hbar\omega$$

Now it can be easily seen, following the above procedure, that

$$[\hat{Q}', \hat{P}'] = i\hbar$$

while

$$[\hat{Q}', \hat{Q}] = \frac{1}{m\omega}\frac{i\hbar e}{c}\left[\frac{\partial A_x}{\partial y} + \frac{\partial A_y}{\partial x}\right] = i\hbar\left[-\frac{B}{2} + \frac{B}{2}\right] = 0$$

$$[\hat{Q}', \hat{P}] = -\frac{i\hbar e}{c}\left[\frac{\partial A_y}{\partial y} + \frac{\partial A_y}{\partial y}\right] = 0$$

where we have used that $\mathbf{A} = \frac{1}{2}(-By, Bx, 0)$ (c.f. problem 17.1). Similarly

$$[\hat{Q}', \hat{P}] = 0 \quad \text{and} \quad [\hat{P}', \hat{P}] = 0.$$

This explains how a seemingly 2-dimensional problem reduces to 1-dimesional oscillator problem. The cyclic character of (Q', P') is reflected in the fact that Landau levels are infinitely degenerate.

Q17.3 For a particle of charge e and mass m, in an electromagnetic field $(\phi, \mathbf{A})$:

$$H = -\frac{\hbar^2}{2m}\nabla^2 + \frac{i\hbar}{mc}e\mathbf{A}\cdot\nabla + \frac{e^2}{2mc^2}\mathbf{A}^2 + e\phi$$

The expectation value of H is

$$\langle\psi|H|\psi\rangle = \langle\psi|H|\psi\rangle^*,$$

Show that

$$2\langle\psi|H|\psi\rangle = \int \left[-\frac{\hbar^2}{2m}(\psi^*\nabla^2\psi + \psi\nabla^2\psi^*) \right.$$

$$+ \left(\frac{ie\hbar}{mc}\right)(\psi^*(\mathbf{A}\cdot\nabla)\psi - \psi(\mathbf{A}\cdot\nabla)\psi^*)$$

$$\left. + \frac{e^2}{mc^2}\psi^*\psi\mathbf{A}^2 + 2e\psi^*\psi\phi \right] d^3x.$$

Using the relation

$$\nabla\cdot(\nabla\psi^*\psi) = \psi\nabla^2\psi^* + \psi^*\nabla^2\psi + 2\nabla\psi^*\cdot\nabla\psi,$$

and the fact that by Gauss's theorem, the left hand side can be written as surface integral for $\nabla(\psi^*\psi)$, which vanishes for a large surface, one gets

$$\langle\psi|H|\psi\rangle = \int d^3x \left[\frac{\hbar^2}{2m}\nabla\psi^*\cdot\nabla\psi + \frac{ie\hbar}{2mc}(\psi^*(\mathbf{A}\cdot\nabla)\psi \right.$$

$$\left. - \psi(\mathbf{A}\cdot\nabla)\psi^*) + \frac{e^2}{2mc^2}\psi^*\psi\mathbf{A}^2 + e\psi^*\psi\phi \right].$$

Hence the Hamiltonian density

$$\mathcal{H} = \frac{\hbar^2}{2m}\nabla\psi^*\cdot\nabla\psi + \frac{ie\hbar}{2mc}(\psi^*\mathbf{A}\cdot\nabla\psi - \psi\mathbf{A}\cdot\nabla\psi^*)$$

$$+ \frac{e^2}{2mc^2}\psi\psi^*\mathbf{A}^2 - e\psi^*\psi\phi$$

$$= -\frac{\hbar^2}{2m}\nabla\psi^*\cdot\nabla\psi + e(\rho\phi - \mathbf{j}\cdot\mathbf{A})$$

where

$$\rho = \psi^*\psi, \quad \mathbf{j} = \frac{\hbar}{2mci}(\psi^*\nabla\psi - \psi\nabla\psi^*) - \frac{e}{2mc^2}\psi^*\psi\mathbf{A}$$

The Lagrangian density $\mathcal{L}_{int} = -\mathcal{H}_{int}$. Verify that

$$\frac{\partial\rho}{\partial t} + \nabla\cdot\mathbf{j} = 0.$$

Solution:

$$\langle\psi|H|\psi\rangle = \int \psi^* \left[-\frac{\hbar^2}{2m}\nabla^2 + \frac{i\hbar}{mc}e\mathbf{A}\cdot\nabla\right.$$

$$\left. + \frac{e^2}{2mc^2}\mathbf{A}^2 + e\phi\right]\psi d^3x$$

$$\langle\psi|H|\psi\rangle^* = \int \psi \left[-\frac{\hbar^2}{2m}\nabla^2 - \frac{i\hbar}{mc}e\mathbf{A}\cdot\nabla\right.$$

$$\left. + \frac{e^2}{2mc^2}\mathbf{A}^2 + e\phi\right]\psi^* d^3x$$

But

$$\langle\psi|H|\psi\rangle^* = \langle\psi|H|\psi\rangle, \quad \text{since } H \text{ is hermitian.}$$

Thus

$$2\langle\psi|H|\psi\rangle = \int \left[-\frac{\hbar^2}{2m}(\psi^*\nabla^2\psi + \psi\nabla^2\psi^*)\right.$$

$$+ \frac{i\hbar}{mc}e\mathbf{A}\cdot(\psi^*\nabla\psi - \psi\nabla\psi^*)$$

$$\left. + \frac{e^2}{mc^2}\psi^*\psi\mathbf{A}^2 + 2e\psi^*\psi\phi\right]d^3x$$

Using

$$\nabla\cdot(\nabla\psi^*\psi) = \nabla\cdot[\psi(\nabla\psi^*) + \psi^*(\nabla\psi)]$$

$$= \psi\nabla^2\psi^* + \psi^*\nabla^2\psi + 2\nabla\psi^*\cdot\nabla\psi,$$

$$\int [\psi^*\nabla^2\psi + \psi\nabla^2\psi^*]d^3x$$

$$= \int \left[\boldsymbol{\nabla} \cdot (\boldsymbol{\nabla}\psi^*\psi) \right] d^3x - 2 \int \left[\boldsymbol{\nabla}\psi^* \cdot \boldsymbol{\nabla}\psi \right] d^3x$$

$$= -2 \int \left(\boldsymbol{\nabla}\psi^* \cdot \boldsymbol{\nabla}\psi \right)$$

where the first integral can be put equal to zero, the integral having total divergence.
Hence

$$\langle \psi | H | \psi \rangle = \int \left[\frac{\hbar^2}{2m} \boldsymbol{\nabla}\psi^* \cdot \boldsymbol{\nabla}\psi + \frac{i\hbar e}{2mc} \mathbf{A} \cdot (\psi^* \boldsymbol{\nabla}\psi - \psi \boldsymbol{\nabla}\psi^*) \right.$$

$$\left. + \frac{e^2}{2mc^2} \psi^* \psi \mathbf{A}^2 + e\psi^* \psi \phi \right] d^3x$$

Hence the Hamiltonian density

$$\mathcal{H} = \frac{\hbar^2}{2m} \boldsymbol{\nabla}\psi^* \cdot \boldsymbol{\nabla}\psi + \frac{ie\hbar}{2mc} \mathbf{A} \cdot (\psi^* \boldsymbol{\nabla}\psi - \psi \boldsymbol{\nabla}\psi^*)$$

$$+ \frac{e^2}{2mc^2} \mathbf{A}^2 \psi^* \psi + e\psi^* \psi \phi$$

$$= \frac{\hbar^2}{2m} \boldsymbol{\nabla}\psi^* \cdot \boldsymbol{\nabla}\psi + e(\rho\phi - \mathbf{J} \cdot \mathbf{A})$$

where

$$\rho = \psi^* \psi, \quad \mathbf{J} = \frac{\hbar}{2mci} (\psi^* \boldsymbol{\nabla}\psi - \psi \boldsymbol{\nabla}\psi^*) - \frac{e}{2mc^2} \psi^* \psi \mathbf{A}$$

Now the Schrödinger equation for the Hamiltonian given in the problem

$$i\hbar \frac{\partial \psi}{\partial t} = \left[-\frac{\hbar^2}{2m} \nabla^2 + \frac{i\hbar e}{mc} \mathbf{A} \cdot \boldsymbol{\nabla} + \frac{e^2}{2mc^2} \mathbf{A}^2 + e\phi \right] \psi$$

$$-i\hbar \frac{\partial \psi^*}{\partial t} = \left[-\frac{\hbar^2}{2m} \nabla^2 - \frac{i\hbar e}{mc} \mathbf{A} \cdot \boldsymbol{\nabla} + \frac{e^2}{2mc^2} \mathbf{A}^2 + e\phi \right] \psi^*$$

Thus

$$\frac{\partial \rho}{\partial t} = \left(\psi^* \frac{\partial \psi}{\partial t} + \psi \frac{\partial \psi^*}{\partial t} \right)$$

$$= \frac{1}{i\hbar} \left(-\frac{\hbar^2}{2m} \right) [\psi^* \nabla^2 \psi - \psi \nabla^2 \psi^*]$$

$$+ \frac{e}{mc} \mathbf{A} \cdot (\psi^* \boldsymbol{\nabla} \psi + \psi \boldsymbol{\nabla} \psi^*) \qquad (17.9)$$

Now

$$\psi^* \nabla^2 \psi - \psi \nabla^2 \psi^* = \boldsymbol{\nabla} \cdot (\psi^* \boldsymbol{\nabla} \psi - \psi \boldsymbol{\nabla} \psi^*)$$

and

$$\mathbf{A} \cdot (\psi^* \boldsymbol{\nabla} \psi + \psi \boldsymbol{\nabla} \psi^*) = \boldsymbol{\nabla} \cdot (\mathbf{A} \psi^* \psi) - \psi^* \psi \boldsymbol{\nabla} \cdot \mathbf{A}$$

$$= \boldsymbol{\nabla} \cdot (\mathbf{A} \psi^* \psi)$$

since in the radiation gauge $\boldsymbol{\nabla} \cdot \mathbf{A} = 0$
$\therefore$ From Eq. (17.9)

$$\frac{\partial \rho}{\partial t} = -\frac{\hbar}{2mi} \boldsymbol{\nabla} \cdot (\psi^* \boldsymbol{\nabla} \psi - \psi \boldsymbol{\nabla} \psi^*)$$

$$+ \frac{e}{mc} \boldsymbol{\nabla} \cdot \mathbf{A} \psi^* \psi$$

$$= -\boldsymbol{\nabla} \cdot \mathbf{J}$$

or

$$\frac{\partial \rho}{\partial t} + \boldsymbol{\nabla} \cdot \mathbf{J} = 0$$

Q17.4 Show that for a hydrogen-like atom, parity operator $\hat{P}$ gives

$$\langle n_f l_f m_f | \mathbf{r} | n_i l_i m_i \rangle = (-1)^{l_i + l_f - 1} \langle n_f l_f m_f | \mathbf{r} | n_i l_i m_i \rangle.$$

Hence parity conservation implies that $(-1)^{l_i} \neq (-1)^{l_f}$ i.e. atomic state must change parity for E_1 transition. This is the parity selection rule. Note also that $(l_f - l_i)$ must be odd.

Solution:

$$\langle n_f l_f m_f | \mathbf{r} | n_i l_i m_i \rangle = \langle n_f l_f m_f | \hat{P}^{-1} \hat{P} \mathbf{r} \hat{P}^{-1} \hat{P} | n_i l_i m_i \rangle$$
$$= -(-1)^{l_f}(-1)^{l_i} \langle n_f l_f m_f | \mathbf{r} | n_i l_i m_i \rangle$$

since

$$\hat{P} \mathbf{r} \hat{P}^{-1} = -\mathbf{r}$$

Thus parity conservation in electromagnetic interaction gives

$$1 = (-1)^{l_f + l_i - 1}$$

This implies

$$l_f \neq l_i, \quad l_f = l_i \pm 1$$

or more generally $l_f - l_i =$ odd number.

Q17.5 By using the relation

$$[L^2, [L^2, \mathbf{r}]] = 2\hbar^2 (\mathbf{r} L^2 + L^2 \mathbf{r}),$$

show that for a hydrogen-like atom

$$l_f - l_i = \pm 1$$

or $l_f = -l_i$. However $l_f = -l_i$ only if $l_f = 0 = l_i$ but then

$$\langle n_f 00 | \mathbf{r} | n_i 00 \rangle.$$

vanishes.

Solution:

$$[L^2, [L^2, \mathbf{r}]] = 2\hbar(\mathbf{r} L^2 + L^2 \mathbf{r})$$

Thus

$$\langle n_f l_f m_f | [L^2, [L^2, \mathbf{r}]] | n_i l_i m_i \rangle$$
$$= 2\hbar^2 \langle n_f l_f m_f | (\mathbf{r} L^2 + L^2 \mathbf{r}) | n_i l_i m_i \rangle$$
$$= 2\hbar^4 [l_f(l_f + 1) + l_i(l_i + 1)]$$
$$\times \langle n_f l_f m_f | \mathbf{r} | n_i l_i m_i \rangle \tag{17.10}$$

Now

$$\text{L.H.S} = \hbar^2 \left[l_f(l_f+1) - l_i(l_i+1)\right] \langle n_f l_f m_f | [L^2, \mathbf{r}] | n_i l_i m_i \rangle$$
$$= \hbar^4 \left[l_f(l_f+1) - l_i(l_i+1)\right]^2 \langle n_f l_f m_f | \mathbf{r} | n_i l_i m_i \rangle$$

Therefore for

$$\langle n_f l_f m_f | \mathbf{r} | n_i l_i m_i \rangle \neq 0$$

we have

$$\left[l_f(l_f+1) - l_i(l_i+1)\right]^2 - 2\left[l_f(l_f+1) + l_i(l_i+1)\right] = 0$$

or

$$(l_f - l_i)^2 \left[(l_f+l_i)^2 + 2(l_f+l_i) + 1\right] - 2(l_f+l_i) - 2(l_f^2+l_i^2) = 0$$

or

$$(l_f + l_i)\left[l_f + l_i + 2\right]\left[(l_f - l_i)^2 - 1\right] = 0$$

i.e. we have two solutions,

$$l_f = -l_i \quad \text{or} \quad l_f - l_i = \pm 1$$

But $l_f = -l_i$ only if $l_f = 0 = l_i$ and in this case

$$\langle n_f 00 | \mathbf{r} | n_i 00 \rangle = 0$$

Hence we have the selection rule

$$\Delta l = \pm 1$$

Q17.6 Photoelectric Effect: Consider the ejection of an electron from the ground state of hydrogen like atom by absorption of a single photon viz. the reaction

$$\text{Photon} + \text{Atom} = e + \text{Atom}'.$$

Calculate the cross-section $\frac{d\sigma}{d\Omega}$ for this process by neglecting the recoil of atom.

Hint: The number of states when the electron has momentum between $\mathbf{p}_e$ and $\mathbf{p}_e + d\mathbf{p}_e = \frac{V}{(2\pi\hbar)^3} d^3 p_e = \frac{V}{(2\pi\hbar)^3} 2m p_e dE_e d\Omega$, $E_e = p_e/2m$. The ground state of hydrogen like atom is represented by the state function

$$\langle \mathbf{r}|a \rangle = \langle r|100 \rangle = u_{100}(r)$$

$$= \frac{1}{\pi}(Z/a_0)^{3/2}e^{-zr/a_0}$$

and for free electron the wave function is given by

$$\langle \mathbf{r}|b \rangle = \frac{1}{\sqrt{V}}e^{i\mathbf{p}_e \cdot \mathbf{r}/\hbar}.$$

Using these results, we can express the total transition probability per unit time for the absorption of photons of frequency ω as

$$W_{ba} = \frac{e^2}{m^2\hbar^2}\frac{2\pi\hbar}{V\omega}\frac{V}{(2\pi\hbar)^3}\int 2mp_e dE_e d\Omega 2\pi\hbar$$

$$\times \delta(E_e + E_B - \hbar\omega)|\langle b|e^{i\mathbf{k}\cdot\mathbf{r}}\boldsymbol{\epsilon}^\lambda\cdot\hat{\mathbf{p}}|a\rangle|^2.$$

Now we can write

$$\langle a|e^{-i\mathbf{k}\cdot\mathbf{r}}\boldsymbol{\epsilon}^\lambda\cdot\hat{\mathbf{p}}|b\rangle = \boldsymbol{\epsilon}^\lambda\cdot\frac{1}{\sqrt{V}}\int d^3 r u_{100}^*(r)e^{-i\mathbf{k}\cdot\mathbf{r}}\mathbf{p}_e e^{i\mathbf{p}_e\cdot\mathbf{r}/\hbar}$$

$$= \frac{1}{\sqrt{V}}\boldsymbol{\epsilon}^\lambda\cdot\mathbf{p}_e\frac{1}{\sqrt{\pi}}(Z/a_0)^{3/2}$$

$$\times \int d^3 r e^{-i(\mathbf{k}-\frac{\mathbf{p}_e}{\hbar})\cdot\mathbf{r}}e^{-zr/a_0}.$$

The flux of incoming photon is $\frac{1}{V}c$ and $\frac{d\sigma}{d\Omega} = \frac{1}{\text{Flux}}\frac{W_{ba}}{d\Omega}$. Note E_B is the binding energy of the electron in the ground state. In evaluating $(\boldsymbol{\epsilon}^\lambda\cdot\mathbf{p}_e)^2$, sum over polarisations; for this purpose it is convenient to select as unit vectors $\boldsymbol{\epsilon}^1$, $\boldsymbol{\epsilon}^2$ and $\frac{\mathbf{k}}{|\mathbf{k}|}$ and take $\mathbf{k}$ along z-axis. The cross-section can be expressed as

$$\frac{d\sigma}{d\Omega} = \frac{32\sqrt{2}Z^5\alpha^8 a^2(E_e/mc^2)^{1/2}\sin^2\theta}{\left\{(\alpha Z)^2 + \frac{2E_e}{mc^2}(1 - v_e/c\cos\theta)\right\}^4},$$

where we have neglected the binding energy and have put $\hbar\omega \approx p_e^2/2m$, $\alpha = e^2/\hbar c$ and v_e is the electron velocity.

Solution: We start from Eq. (17.97c) of the text

$$\sigma_{ba}^{abs}(\omega) = \frac{4\pi^2 e^2}{m^2 \omega c} |\langle b|e^{i\mathbf{k}\cdot\mathbf{r}}\epsilon^\lambda \cdot \hat{\mathbf{p}}|a\rangle|^2 \delta(E_b - E_a - \hbar\omega)$$

$$(17.11)$$

We make the assumption that $\hbar\omega$ is much larger than the ionization potential of the atom so that the final electron states can be approximated by a plane wave (i.e electron is unperturbed by atomic forces in its departure)

$$u_b(\mathbf{r}) = \langle\mathbf{r}|b\rangle = \frac{1}{\sqrt{V}}e^{i\mathbf{p}_b\cdot\mathbf{r}/\hbar}$$

Then using $u_a(\mathbf{r}) = \langle\mathbf{r}|a\rangle$ as given in the problem, we can write

$$\langle b|e^{i\mathbf{k}\cdot\mathbf{r}}\epsilon^\lambda \cdot \hat{\mathbf{p}}|a\rangle$$

$$= \frac{1}{\sqrt{\pi}}\left(\frac{Z}{a_0}\right)^{3/2}\frac{1}{\sqrt{V}}$$

$$\times \int d^3 r\, e^{-i\mathbf{p}_b\cdot\mathbf{r}/\hbar}e^{i\mathbf{k}\cdot\mathbf{r}}\epsilon_\lambda \cdot (-i\hbar\boldsymbol{\nabla})e^{-Zr/a_0}$$

Since $\hat{\mathbf{p}}$ is hermitian

$$\langle b|e^{i\mathbf{k}\cdot\mathbf{r}}\epsilon^\lambda \cdot \hat{\mathbf{p}}|a\rangle^*$$

$$= \langle a|e^{-i\mathbf{k}\cdot\mathbf{r}}\epsilon^\lambda \cdot \hat{\mathbf{p}}|b\rangle$$

$$= \frac{1}{\sqrt{\pi}}\left(\frac{Z}{a_0}\right)^{3/2}\frac{1}{\sqrt{V}}$$

$$\times \int d^3 r\, e^{-Zr/a_0} \times e^{-i\mathbf{k}\cdot\mathbf{r}}(-i\hbar\boldsymbol{\nabla})e^{i\mathbf{p}_b\cdot\mathbf{r}/\hbar}$$

$$= \frac{1}{\sqrt{\pi}}\left(\frac{Z}{a_0}\right)^{3/2}\frac{1}{\sqrt{V}}(-i\hbar)$$

$$\times \int d^3 r\, e^{-Zr/a_0}\epsilon^\lambda \cdot \left(i\frac{\mathbf{p}_b}{\hbar}\right)e^{-i\mathbf{k}\cdot\mathbf{r}}e^{i\mathbf{p}_b\cdot\mathbf{r}/\hbar}$$

$$= \frac{1}{\sqrt{\pi}}\left(\frac{Z}{a_0}\right)^{3/2}\frac{1}{\sqrt{V}}\epsilon^\lambda \cdot \mathbf{p}_b \int d^3 r\, e^{-Zr/a_0}e^{-i\mathbf{q}\cdot\mathbf{r}}$$

$$(17.12)$$

where $\mathbf{q} = \mathbf{k} - \mathbf{p}_b/\hbar$. Now $d^3r = 2\pi r^2 \sin\theta d\theta$. In order to perform the integration, take $\mathbf{q}$ along z-axis, so that $\mathbf{q} \cdot \mathbf{r} = qr\cos\theta$. The integral gives

$$\int d^3 r e^{-Zr/a_0} e^{-i\mathbf{q}\cdot\mathbf{r}} = \frac{8\pi(Z/a_0)}{[q^2 + (Z/a_0)^2]^2}$$

Hence we have

$$\langle b|e^{i\mathbf{k}\cdot\mathbf{r}}\boldsymbol{\epsilon}_\lambda \cdot \hat{\mathbf{p}}|a\rangle^* = \frac{1}{\sqrt{V}}\frac{8\pi}{\sqrt{\pi}}\left(\frac{Z}{a_0}\right)^{5/2}\frac{(\boldsymbol{\epsilon}_\lambda \cdot \mathbf{p}_b)}{[q^2 + (Z/a_0)^2]^2}$$

$$(17.13)$$

Now $|b\rangle$ lies in the continuum, so that

$$\sigma_{abs}(\omega) = \frac{4\pi^2 e^2}{\omega c m^2}\sum_b \left|\langle b|e^{i\mathbf{k}\cdot\mathbf{r}}\boldsymbol{\epsilon}_\lambda \cdot \hat{\mathbf{p}}|a\rangle\right|^2 \delta(E_b - E_a - \hbar\omega)$$

$$(17.14)$$

Now $\mathbf{p}_b = \hbar\mathbf{k}_b$ and

$$\frac{1}{V}\sum_b = \frac{1}{(2\pi\hbar)^3}d\Omega\int p_b^2 dp_b$$

Thus

$$\sum_b \rightarrow \left[\frac{V}{(2\pi\hbar)^3}\right]d\Omega\int p_b^2 dp_b \qquad (17.15)$$

Now

$$E_b = \frac{p_b^2}{2m}, \quad dE_b = \frac{p_b dp_b}{m}$$

so that

$$\int p_b^2 dp_b \delta(E_b - E_a - \hbar\omega) = \int p_b m \delta(E_b - E_a - \hbar\omega)dE_b$$

$$= mp_b \qquad (17.16)$$

where now $(E_b = E_e)$

$$p_b = \sqrt{2mE_e} = \sqrt{2m(E_a + \hbar\omega)} \simeq \sqrt{2m\hbar\omega} \qquad (17.17)$$

If the incident photon is unpolarized, we must average over initial polarization i.e. we have

$$\frac{1}{2}\sum_{\lambda=1}^{b}|\boldsymbol{\epsilon}_\lambda\cdot\mathbf{k}_b|^2 = \frac{1}{2}k_{bi}k_{bj}\left(\delta_{ij} - \frac{k_ik_j}{k^2}\right)$$

$$\frac{1}{2}k_b^2(1 - \cos^2\theta) = \frac{1}{2}k_b^2\sin^2\theta \qquad (17.18)$$

where $\cos\theta = \frac{\mathbf{k}\cdot\mathbf{k}_b}{|\mathbf{k}||\mathbf{k}_b|}$. Further we have to calculate

$$\left(\frac{Z^2}{a_0^2} + q^2\right), \quad \text{where } q^2 = (\mathbf{k}_b - \mathbf{k})^2$$

$$= k_b^2(1 - 2(k/k_b)\cos\theta + (k/k_b)^2) \qquad (17.19)$$

and

$$\frac{k}{k_b} = \frac{\omega}{ck_b} = \frac{\hbar\omega}{\hbar ck_b} = \frac{p_b^2/2m}{cp_b}$$

$$= \frac{v_e}{2c}, \qquad (17.20)$$

where v_e is the electron velocity and

$$k_b^2 = \frac{p_b^2}{\hbar^2} = \frac{2m}{\hbar^2}E_e.$$

Thus finally, neglecting terms of order $O(v_e^2/c^2)$,

$$\left(\frac{Z^2}{a_0} + q^2\right) = \frac{m^2c^2}{\hbar^2}\left(Z^2\left(\frac{e^2}{\hbar c}\right)^2 + \frac{2E_e}{mc^2}\left(1 - \frac{v_e}{c}\cos\theta\right)\right)$$

$$\qquad (17.21)$$

where we have $a_0^2 = \frac{\hbar^4}{m^2e^4}$. Thus from Eqs. (17.14), (17.15), (17.16), (17.17), (17.18) and (17.21)

$$\frac{d\sigma}{d\Omega} = \frac{32e^2\frac{1}{2}\sqrt{\frac{2mE_e}{\hbar^2}}\frac{2mE_e}{\hbar^2}\sin^2\theta Z^5}{mc\frac{E_e}{\hbar}a_0^5\left(\frac{m^2c^2}{\hbar^2}\right)^4\left[Z^2\left(\frac{e^2}{\hbar c}\right)^2 + \frac{2E_e}{mc^2}\left(1 - \frac{v_e}{c}\cos\theta\right)\right]^4}$$

$$= \frac{32\sqrt{2}Z^5\alpha^8 a_0^2(E_e/mc^2)^{1/2}\sin^2\theta}{\left[(Z\alpha)^2 + (2E_e/mc^2)(1 - \frac{v_e}{c}\cos\theta)\right]^4}$$

Q17.7 Show that $\langle 0|E_i(\mathbf{r}, t)E_j(\mathbf{r}', t)|0\rangle$, the expectation value of electric field at 2 points in vacuum (i.e. no photons), is given by

$$4\pi\hbar c^2 \left(\delta_{ij}\mathbf{n}\cdot\mathbf{n}' - \frac{\partial}{\partial x_i}\frac{\partial}{\partial x_j'}\right)D_1(\mathbf{R}),$$

where ($\omega = c|\mathbf{k}|$)

$$D_1(\mathbf{R}) = \frac{1}{(2\pi)^3}\int\frac{d^3k}{2\omega}e^{i\mathbf{k}\cdot(\mathbf{r}-\mathbf{r}')}, \quad \mathbf{R} = \mathbf{r} - \mathbf{r}'$$

$$= \frac{1}{4\pi^2|\mathbf{R}|c}\int_0^\infty dk\sin k|\mathbf{R}|.$$

By defining the last integral as

$$\lim_{\alpha\to 0}\int_0^\infty e^{-\alpha k}\sin k|\mathbf{R}|dk,$$

show that

$$D_1(R) = \frac{1}{4\pi^2 c|\mathbf{R}|^2}$$

and $D_1(R) \to \infty$ when $\mathbf{R} \to 0$, i.e. $\mathbf{r} = \mathbf{r}'$. Since what we measure by a test body is the field strength averaged over some region in space, it may be more realistic to consider the average field operator about some point, e.g. defined as

$$\bar{\mathbf{E}} = \frac{1}{\Delta V}\int_{\Delta V}\mathbf{E}d^3r,$$

where ΔV is a small volume containing the position in question. Then show that

$$\langle 0|\bar{\mathbf{E}}\cdot\bar{\mathbf{E}}|0\rangle \sim \hbar c/(\Delta\ell)^4, \tag{I}$$

where $\Delta\ell$ is the linear dimension of volume ΔV. The above expression characterizes the fluctuations in the electric field when no photons are present. Compare it with the square of the field strength for a classical electromagnetic wave of wavelength $2\pi\lambda$ where the time average of $\mathbf{E}^2$

can be equated with the energy density of electromagnetic wave so that

$$(\mathbf{E}^2)_{\text{average}} = \bar{n}\hbar(c/\lambda), \tag{II}$$

where $\bar{n}$ stands for the number of photons per unit volume. Thus, for the validity of classical description, purely quantum effects such as (I) (with $\Delta\ell \sim \lambda$) must be completely negligible in comparison with (II). For this, we must have

$$\bar{n} \gg \frac{1}{\lambda^3},$$

i.e. the description of physical phenomenon based on classical electrodynamics is reliable when number of photons per volume λ^3 is $\gg 1$. Show that this condition is satisfied by number of photons per volume λ^3 at a distance of 10 km from an antenna with a power of 135,000 Watts giving waves of $\lambda \simeq 50$ cm.

Solution:

$$E_i(\mathbf{r}, t) = -\frac{1}{c}\frac{\partial A_i}{\partial t}$$

$$= i\frac{\sqrt{4\pi\hbar}}{\sqrt{V}}\sum_{\mathbf{k}}\sum_{\lambda}\sqrt{\frac{\omega}{2}}\Big[\epsilon_{\lambda i}a_\lambda(\mathbf{k})e^{i\mathbf{k}\cdot\mathbf{r}}e^{-i\omega t}$$

$$- \epsilon_{\lambda i}^* a_\lambda^\dagger(\mathbf{k})e^{-i\mathbf{k}\cdot\mathbf{r}}e^{i\omega t}\Big] \tag{17.22}$$

$$E_j(\mathbf{r}, t) = i\frac{\sqrt{4\pi\hbar}}{\sqrt{V}}\sum_{\mathbf{k'}}\sum_{\lambda'}\sqrt{\frac{\omega'}{2}}\Big[\epsilon_{\lambda'j}a_{\lambda'}(\mathbf{k'})e^{i\mathbf{k'}\cdot\mathbf{r'}}e^{-i\omega't}$$

$$- \epsilon_{\lambda'j}^* a_{\lambda'}^\dagger(\mathbf{k'})e^{-i\mathbf{k'}\cdot\mathbf{r'}}e^{i\omega't}\Big] \tag{17.23}$$

In calculating $\langle 0|E_i(\mathbf{r}, t)E_j(\mathbf{r'}, t)|0\rangle$, note that $a_{\lambda'}|0\rangle = 0$ and $\langle 0|a_\lambda^\dagger(\mathbf{k}) = 0$. Thus the only term which contributes is the one which involves $a_\lambda(\mathbf{k})a_{\lambda'}^\dagger(\mathbf{k'})$

$$\therefore \langle 0|E_i(\mathbf{r}, t)E_j(\mathbf{r'}, t)|0\rangle = i^2(-1)\frac{4\pi\hbar}{V}\sum_{\mathbf{k}}\sum_{\mathbf{k'}}\sum_{\lambda}\sum_{\lambda'}\frac{\sqrt{\omega\omega'}}{2}$$

$$\times \langle 0|\epsilon_{\lambda i}(\mathbf{k})e^{i\mathbf{k}\cdot\mathbf{r}}e^{-i\omega t}a_\lambda(\mathbf{k})\epsilon_{\lambda'j}^*(\mathbf{k'})$$

$$\times e^{-i\mathbf{k'}\cdot\mathbf{r'}}e^{i\omega t}a_{\lambda'}^\dagger(\mathbf{k'})|0\rangle \tag{17.24}$$

Now using the commutator

$$[a_\lambda(\mathbf{k}), a_{\lambda'}^\dagger(\mathbf{k}')] = \delta_{\lambda\lambda'}\delta_{\mathbf{k}\mathbf{k}'}$$

$$a_\lambda(\mathbf{k})a_{\lambda'}^\dagger(\mathbf{k}')|0\rangle = \delta_{\lambda\lambda'}\delta_{\mathbf{k}\mathbf{k}'}|0\rangle + a_{\lambda'}^\dagger(\mathbf{k}')a_\lambda(\mathbf{k})|0\rangle$$

$$= \delta_{\lambda\lambda'}\delta_{\mathbf{k}\mathbf{k}'}|0\rangle$$

Further $\delta_{\lambda\lambda'}\delta_{\mathbf{k}\mathbf{k}'}$ remove summations over $\sum_{\mathbf{k}'}\sum_{\lambda'}$ in Eq. (17.24) and we obtain

$$\langle 0|E_i(\mathbf{r}, t)E_j(\mathbf{r}', t)|0\rangle$$

$$= \frac{4\pi\hbar}{V}\sum_{\mathbf{k}}\left(\sum_\lambda \epsilon_{\lambda i}(\mathbf{k})\epsilon_{\lambda j}^*(\mathbf{k}')\right)\frac{\omega}{2}e^{i\mathbf{k}\cdot(\mathbf{r}-\mathbf{r}')} \quad (17.25)$$

Now

$$\sum_\lambda \epsilon_{\lambda i}(\mathbf{k})\epsilon_{\lambda j}^*(\mathbf{k}) = \delta_{ij} - \frac{k_i k_j}{k^2} \quad (17.26)$$

$$\frac{1}{V}\sum_{\mathbf{k}} \to \frac{1}{(2\pi)^3}\int d^3 k \quad (17.27)$$

Thus

$$\langle 0|E_i(\mathbf{r}, t)E_j(\mathbf{r}', t)|0\rangle$$

$$= 4\pi\hbar\frac{1}{(2\pi)^3}\int \frac{d^3 k}{2\omega}\omega^2 \underbrace{\left(\delta_{ij} - \frac{k_i k_j}{k^2}\right)}_{i^2\delta_{ij}\boldsymbol{\nabla}\cdot\boldsymbol{\nabla}' - c^2\frac{\partial}{\partial x_i}\frac{\partial}{\partial x_j'}} e^{i\mathbf{k}\cdot(\mathbf{r}-\mathbf{r}')}$$

$$\times 4\pi\hbar c^2\left[\delta_{ij}\boldsymbol{\nabla}\cdot\boldsymbol{\nabla}' - \frac{\partial}{\partial x_i}\frac{\partial}{\partial x_j'}\right]D_1(\mathbf{R})$$

where $\omega = ck$ and

$$D_1(\mathbf{R}) = \frac{1}{(2\pi)^3}\int \frac{d^3 k}{2\omega}e^{i\mathbf{k}\cdot\mathbf{R}}, \quad \mathbf{R} = \mathbf{r} - \mathbf{r}'$$

Now

$$d^3 k = k^2 dk\, 2\pi \sin\theta d\theta$$

Thus

$$D_1(\mathbf{R}) = \frac{2\pi}{(2\pi)^3} \int_0^\infty \frac{k^2 dk}{2ck} \int_0^\pi e^{ik|\mathbf{R}|\cos\theta} \sin\theta d\theta$$

$$= \frac{1}{4\pi^2 c|\mathbf{R}|} \int_0^\infty dk \sin k|\mathbf{R}|$$

To give a meaning to the integral we write it as

$$\lim_{\alpha \to 0} \int e^{-\alpha k} \sin k|\mathbf{R}| dk$$

$$= \lim_{\alpha \to 0} \left[e^{-\alpha k} \frac{-\cos k|\mathbf{R}|}{|\mathbf{R}|} \Big|_0^\infty - \int_0^\infty \frac{\cos k|\mathbf{k}|}{|\mathbf{R}|} (-\alpha) e^{-\alpha k} \right]$$

$$= \frac{1}{|\mathbf{R}|}$$

Thus

$$D_1(\mathbf{R}) = \frac{1}{4\pi^2 c|\mathbf{R}|^2}$$

$D_1(R) \to \infty$ when $R \to 0$ i.e. $\mathbf{r} = \mathbf{r}'$

Now

$$\overline{E}_i = \frac{1}{\Delta V} \int_{\Delta V} E_i(\mathbf{r}, t) d^3 r$$

$$\overline{E}_j = \frac{1}{\Delta V} \int_{\Delta V'} E_j(\mathbf{r}', t) d^3 r'$$

Thus

$$\langle 0|\overline{\mathbf{E}}^2|0\rangle = \frac{1}{(\Delta V)^2} \int_{\Delta V} \int_{\Delta V'} d^3 r d^3 r'$$

$$\times \langle 0|E_i(\mathbf{r}, t) E_i(\mathbf{r}', t)|0\rangle$$

$$= \frac{1}{(\Delta V)^2} \int_{\Delta V} \int_{\Delta V'} 4\pi \hbar c^2 [3\boldsymbol{\nabla} \cdot \boldsymbol{\nabla}']$$

$$\times \frac{1}{4\pi^2 c} \frac{1}{|\mathbf{R}|^2} d^3 r d^3 r'$$

$$= \frac{2c\hbar}{\pi} \frac{1}{(\Delta V)^2} \int_{\Delta V} \int d^3 r \boldsymbol{\nabla} \cdot \underbrace{\int d^3 r' \boldsymbol{\nabla}' \frac{1}{|\mathbf{R}|^2}}_{\int_{\Delta S'} \frac{1}{|\mathbf{R}|^2} d\mathbf{S}'}$$

By using Gauss's law as indicated

$$\int d^3 r' \boldsymbol{\nabla}' \frac{1}{|R|^2} \simeq \frac{\hat{r}}{r^2}(\Delta l)^2 \qquad (17.28)$$

we obtain

$$\langle 0|\overline{\mathbf{E}}^2|0\rangle = \frac{2c\hbar}{\pi}\frac{(\Delta l)^2}{(\Delta l)^6}\int_{\Delta V} d^3 r \boldsymbol{\nabla} \cdot \left(\frac{\hat{r}}{r^2}\right)$$

Again by using Gauss's law

$$\int_{\Delta V} d^3 r \boldsymbol{\nabla} \cdot \left(\frac{\hat{r}}{r^2}\right) \simeq \int_{dS} \frac{\hat{r} \cdot d\mathbf{S}}{r^2} \simeq \frac{(\Delta l)^2}{(\Delta l)^2} \approx 1$$

and thus

$$\langle 0|\overline{\mathbf{E}}^2|0\rangle = \frac{2}{\pi}c\hbar\frac{1}{(\Delta l)^4} \approx \frac{c\hbar}{(\Delta l)^4}$$

Now

$$u_{av} \equiv (\overline{\mathbf{E}}^2) = \frac{S_{av}}{c} \qquad \text{where } \mathbf{S} \text{ is the Poynting vector}$$

$$= \frac{1.35 \times 10^5 \frac{\text{Watts}}{4\pi R^2}}{c}, \qquad R = 10\,\text{km} = 10^4\,\text{m}$$

$$= \frac{1.35 \times 10^5 \text{Watts}}{4\pi 10^8\,\text{m}^2 \cdot 3 \times 10^8\,\text{m/s}}$$

$$= \frac{1.35}{12\pi}10^{-11}\frac{\text{J}}{\text{m}^3} \qquad \text{J is for Joule}$$

$$= \bar{n}\hbar(c/\lambda)$$

$$= \bar{n}(1.06 \times 10^{-34}\,\text{J-s})\frac{3 \times 10^8\,\text{m/s}}{\lambda},$$

$$\lambda = 50\,\text{cm} = \frac{1}{2}\,\text{m}$$

$$\bar{n} \simeq \frac{1.35}{12\pi}\frac{10^{-11} \times 10^{26}}{6}(1/\text{m}^3)$$

$$\simeq \frac{10^{13}}{\text{m}^3} \gg (1/\lambda^3)$$

Q17.8 For quantised radiation field, show that

$$[E_i(\mathbf{r}, t), B_j(\mathbf{r}', t)] = -ic(4\pi)\hbar\epsilon_{ijk}\frac{\partial}{\partial x_k}\delta(\mathbf{r} - \mathbf{r}').$$

What does this say about the simultaneous measureability of **B** and **E**?

Solution: $E_i(\mathbf{r}, t)$ is given in Eq. (1) of Problem (17.6)

$$B_j(\mathbf{r}, t) = (\nabla \times \mathbf{A})$$

$$= \epsilon_{jln}\frac{\sqrt{4\pi\hbar c}}{\sqrt{V}}\sum_{k'}\sum_{\lambda'}\frac{1}{\sqrt{2\omega'}}ik_l'\epsilon_{\lambda'm}a_{\lambda'}(k')$$

$$\times e^{ik'\cdot\mathbf{r}}e^{-i\omega't} - ik_{l'}\epsilon_{\lambda'm}^*a_{\lambda'}^\dagger(k')$$

$$\times e^{-ik'\cdot\mathbf{r}'}e^{i\omega't} \tag{17.29}$$

$$[E_i(\mathbf{r}, t), B_j(\mathbf{r}', t)] = \epsilon_{jlm}\frac{4\pi\hbar c^2}{V}\frac{i}{c}\sum_k\sum_{k'}\sum_\lambda\sum_{\lambda'}\frac{\omega}{2\sqrt{\omega\omega'}}$$

$$\times\epsilon_{\lambda i}(k)\epsilon_{\lambda'm}^*(k')(-ik_l')[a_\lambda(k), a_{\lambda'}^\dagger(k')]$$

$$\times e^{ik\cdot\mathbf{r}}e^{-ik'\cdot\mathbf{r}'}$$

$$+(-ik_l')[a_\lambda^\dagger(k), a_{\lambda'}(k')]\epsilon_{\lambda i}^*(k)\epsilon_{\lambda'm}^*(k')$$

$$\times e^{-ik\cdot\mathbf{r}}e^{ik'\cdot\mathbf{r}'}$$

Now using

$$[a_\lambda(k), a_{\lambda'}^\dagger(k')] = \delta_{\lambda\lambda'}\delta_{kk'}$$

the summation over $\sum_{\lambda'}\sum_{k'}$ is removed and further

$$\sum_\lambda\epsilon_{\lambda i}(k)\epsilon_{\lambda m}^*(k) = \sum_\lambda\left(\delta_{im} - \frac{k_ik_m}{k^2}\right)$$

$$= \sum_\lambda\epsilon_{\lambda i}^*(k)\epsilon_{\lambda m}(k)$$

Then from (17.29)

$$[E_i(\mathbf{r}, t), B_j(\mathbf{r}', t)]$$

$$= \frac{1}{2} \frac{4\pi \hbar c}{V} \epsilon_{jlm} i \sum_k \left(\delta_{im} - \frac{k_i k_m}{k^2} \right)$$

$$\times [-ik_l e^{i\mathbf{k} \cdot (\mathbf{r} - \mathbf{r}')} + ik_l e^{-i\mathbf{k} \cdot (\mathbf{r} - \mathbf{r}')}]$$

$$= \frac{4\pi \hbar c}{V} \epsilon_{jli} \sum_k \frac{1}{2} k_l (e^{i\mathbf{k} \cdot (\mathbf{r} - \mathbf{r}')} - e^{-i\mathbf{k} \cdot (\mathbf{r} - \mathbf{r}')})$$

Now $\frac{1}{V} \sum_k \to \frac{1}{(2\pi)^3} \int d^3k$ and changing $\mathbf{k} \to -\mathbf{k}$ in the second integral and using

$$\int d^3k \, e^{i\mathbf{k} \cdot (\mathbf{r} - \mathbf{r}')} = (2\pi)^3 \delta(\mathbf{r} - \mathbf{r}')$$

we get

$$[E_i(\mathbf{r}, t), B_j(\mathbf{r}', t)] = 4\pi \hbar c \epsilon_{jli} \left(-i \frac{\partial}{\partial x_l} \right) \delta(\mathbf{r} - \mathbf{r}')$$

This implies that we cannot simultaneously measure E_i and B_j at the same point in space.

Additional Problem:

Q1 Given the Lagrangian density

$$\mathcal{L} = -\frac{\hbar^2}{2m} \boldsymbol{\nabla} \psi^* \cdot \boldsymbol{\nabla} \psi - \frac{i\hbar}{2} \left(\psi \frac{\partial \psi^*}{\partial t} - \psi^* \frac{\partial \psi}{\partial t} \right),$$

derive the interaction Lagrangian density $\mathcal{L}_{int}$ for the electromagnetic field $(\phi, \mathbf{A})$ by replacing the derivatives by the corresponding co-variant derivatives:

$$\boldsymbol{\nabla} \to \boldsymbol{\nabla} - \frac{ie}{c\hbar} \hat{A}; \quad \frac{\partial}{\partial t} \to \frac{\partial}{\partial t} + \frac{ie}{\hbar} \phi$$

Solution:

$$\mathcal{L} = -\frac{\hbar^2}{2m} \left[\left(\boldsymbol{\nabla} \psi^* + \frac{ie}{c\hbar} \psi^* \mathbf{A} \right) \cdot \left(\boldsymbol{\nabla} \psi - \frac{ie}{c\hbar} \psi \mathbf{A} \right) \right]$$

$$- \frac{i\hbar}{2} \left[\psi \left(\frac{\partial \psi^*}{\partial t} - \frac{ie}{\hbar} \phi \psi^* \right) - \psi^* \left(\frac{\partial \psi}{\partial t} + \frac{ie}{c\hbar} \phi \psi \right) \right]$$

$$= -\frac{\hbar^2}{2m}\boldsymbol{\nabla}\psi^* \cdot \boldsymbol{\nabla}\psi - \frac{i\hbar}{2}\left(\psi\frac{\partial\psi^*}{\partial t} - \psi^*\frac{\partial\psi}{\partial t}\right)$$

$$- \frac{ie\hbar}{2mc}(\psi^*\boldsymbol{\nabla}\psi - \psi\boldsymbol{\nabla}\psi^*)\cdot\mathbf{A}$$

$$- \frac{e^2}{2mc^2}\psi^*\psi\mathbf{A}^2 - e\bar{\psi}\psi\phi$$

Hence

$$\mathcal{L}_{int} = e(\mathbf{j}\cdot\mathbf{A} - \rho\phi) = -\mathcal{H}_{int}$$

where

$$\mathbf{j} = \left[\frac{\hbar}{2imc}(\psi^*\boldsymbol{\nabla}\psi - \psi\boldsymbol{\nabla}\psi^*) - \frac{e^2}{2mc^2}\psi^*\psi\mathbf{A}\right]$$

$$\rho = \psi^*\psi$$

Chapter 18

Formal Theory of Scattering

Q18.1 A system is prepared in the state $|a'\rangle$ at $t = 0$. Show that the probability of finding it in the state $|b'\rangle$ at time t is given by

$$|\langle b'|e^{-iH_0t/\hbar}U(t,0)|a'\rangle|^2,$$

where U is the time displacement operator in the interaction picture

$$|a(t)\rangle_I = U(t,t_0)|a(t_0)\rangle_I.$$

Solution:

$$c_{b'}(t) = \langle b'|a(t)\rangle$$
$$= \langle b'|e^{-iH_0t/\hbar}|a(t)\rangle_I = \langle b'|e^{-iH_0t/\hbar}U(t,0)|a(0)\rangle_I$$

But $|a(0)\rangle_I = |a'\rangle$, using Eq. (18.13) of the text for $t = 0$. Hence the required probability is

$$|c_{b'}(t)|^2 = |\langle b'|e^{-iH_0t/\hbar}U(t,0)|a'\rangle|^2$$

Q18.2 Show that the unitarity of S-matrix follows from the conservation of probability.

Solution:

$$c_b(t) = \langle b|U(t,-\infty)|a\rangle$$

so that

$$|c_b(t)|^2 = |\langle b|U(t,-\infty)|a\rangle|^2$$
$$= \langle b|U(t,-\infty)|a\rangle\langle b|U(t,-\infty)|a\rangle^*$$
$$= \langle a|U^\dagger(t,-\infty)|b\rangle\langle b|U(t,-\infty)|a\rangle$$

Thus $([\sum_b |b\rangle\langle b|] = \hat{I})$

$$\sum_b |c_b(t)|^2 = \sum_b \langle a|U^\dagger(t,-\infty)|b\rangle\langle b|U(t,-\infty)|a\rangle$$
$$= \langle a|U^\dagger(t,-\infty)U(t,-\infty)|a\rangle$$
$$= \langle a|U(-\infty,t)U(t,-\infty)|a\rangle$$
$$= \langle a|U(-\infty,-\infty)|a\rangle$$
$$= \langle a|S^\dagger S|a\rangle, \quad \text{since } S = U(\infty,-\infty),$$
$$S^\dagger = U(-\infty,\infty) \tag{18.1}$$

Conservation of probability gives

$$\sum_b |c_b(t)|^2 = 1 \quad \text{which implies } S^\dagger S = 1$$

Q18.3 If the S-matrix is represented as

$$S = \frac{1 - i/2\ K}{1 + i/2\ K},$$

where K is a hermitian operator, show that it satisfies $SS^\dagger = 1$. Derive the relation (analogue of unitarity relation)

$$T_{fi} = -K_{fi} - i\pi \sum_a K_{fa}T_{ai}\delta(E_a - E_i),$$

where

$$\langle f|K|i\rangle = 2\pi\delta(E_f - E_i)K_{fi}.$$

Solution:

$$S = \frac{1 - \frac{i}{2}K}{1 + \frac{i}{2}K} \qquad K^\dagger = K$$

$$S^\dagger = \frac{1 + \frac{i}{2}K}{1 - \frac{i}{2}K} \qquad SS^\dagger = 1$$

$$\left(1 + \frac{i}{2}K\right)S = 1 - \frac{i}{2}K$$

This gives

$$\sum_a \langle f|\left(1 + \frac{i}{2}K\right)|a\rangle\langle a|S|i\rangle = \delta_{fi} - \frac{i}{2}\langle f|K|i\rangle$$

Now L.H.S

$$\sum_a \left[\delta_{fa} + \frac{i}{2}\langle f|K|a\rangle\right]\left[\delta_{ai} + 2\pi i\delta(E_a - E_i)T_{ai}\right]$$

$$= \delta_{fi} + 2\pi i\delta(E_f - E_i)T_{fi} + \frac{i}{2}\langle f|K|i\rangle$$

$$- \pi \sum_a \delta(E_a - E_i)\langle f|K|a\rangle T_{ai}$$

Hence we have

$$2\pi i\delta(E_i - E_f)T_{fi} = -i\langle f|K|i\rangle + \pi \sum_a \delta(E_a - E_i)$$

$$\times \langle f|K|a\rangle T_{ai}$$

On using

$$\langle f|K|i\rangle = 2\pi\delta(E_f - E_i)K_{fi},$$

we get

$$2\pi i\delta(E_f - E_i)T_{fi} = -2\pi i\delta(E_f - E_i)K_{fi}$$

$$+ 2\pi^2 \sum_a \delta(E_a - E_i)\delta$$

$$\times (E_f - E_a)K_{fa}T_{ai}$$

$$= -2\pi i \delta(E_f - E_i) K_{fi}$$
$$- i(2\pi i)\pi \sum_a \delta(E_a - E_i)$$
$$\times \delta(E_f - E_a) K_{fa} T_{ai}$$

Hence

$$T_{fi} = -K_{fi} - i\pi \sum_a K_{fa} T_{ai} \delta(E_a - E_i)$$

Q18.4 Show that for the scattering process

$$\alpha + Y_0 \rightarrow \alpha + Y_n,$$

if the potential $U(r)$, which possesses the bound states Y_n, is central, the inelastic form factor $(n \neq 0)$ given in Eq. ((18.99b) of the text) is

$$F_{lm}(q) = 4\pi i^l Y_{lm}^*(\hat{\mathbf{q}}) \int_0^\infty u_n^*(r) j_l(qr) u_0(r) r^2 dr,$$

where $l(l+1)\hbar^2$ and $m\hbar$ are eigenvalues of L^2 and L_z. Show further that
(i) $F_{lm}(q) \sim q^l$ as $q \rightarrow 0$, (ii) the partial wave cross-section is

$$\frac{d\sigma_l}{d\Omega_\alpha'} = (2l+1)\frac{p_\alpha'}{p_\alpha}\sigma^B(q) \left| \int_0^\infty u_n^*(r) j_l(qr) u_0(r) r^2 dr \right|^2,$$

where $u_n(r)$ are the solutions of the radial Schrödinger equation.
Hint: Write

$$\phi_n(\mathbf{r}) = \langle \mathbf{r}|n \rangle = Y_{lm}(\hat{\mathbf{r}}) u_n(r)$$
$$e^{i\mathbf{q}\cdot\mathbf{r}} = 4\pi \sum_{l'm'} i^{l'} j_{l'}(qr) Y_{l'm'}^*(\hat{\mathbf{q}}) Y_{l'm'}(\hat{\mathbf{r}})$$

and note

$$j_l(qr) \sim (qr)^l \quad \text{as} \quad q \rightarrow 0.$$

In calculating the partial wave cross-section, remember that we do not distinguish final states which only differ in magnetic quantum number m so that the observed cross-section involves the sum

$$\sum_m |Y_{lm}(\hat{q})|^2 = \frac{2l+1}{4\pi}.$$

Solution: Using the expressions for $\phi_n(\mathbf{r})$ and $e^{i\mathbf{q}\cdot\mathbf{r}}$ given above, the form factor $F_n(q)$ given in Eq. (18.99b) of the text can be written as

$$F_{lm}(q) = \frac{4\pi}{\sqrt{4\pi}} \sum_{l'm'} i^l Y_{l'm'}(\hat{q}) \left[\int Y_{lm}^*(\hat{\mathbf{r}}) Y_{l'm'}(\hat{\mathbf{r}}) d\Omega \right]$$

$$\times \int_0^\infty u_n^*(r) j_l(qr) u_0(r) r^2 dr$$

By using the orthogonality of Y's

$$\int Y_{lm}^*(\hat{\mathbf{r}}) Y_{l'm'}(\hat{\mathbf{r}}) d\Omega = \delta_{ll'} \delta_{mm'}$$

we have

$$F_{lm}(q) = \sqrt{4\pi} i^l Y_{lm}^*(\hat{q}) \int_0^\infty u_n^*(r) j_l(qr) u_0(r) r^2 dt$$

Now

$$j_l(qr) \sim (qr)^l, \qquad q \to 0$$
$$F_{lm}(q) \sim q^l, \qquad \text{as } q \to 0$$

In practice we do not distinguish final states which differ only in magnetic quantum number m so that the observed cross-section involves the sum

$$\sum_m |Y_{lm}(\hat{q})|^2 = \frac{2l+1}{4\pi}$$

Thus from Eq. (18.101) of the text

$$\frac{d\sigma_l}{d\Omega_{\alpha'}} = (2l+1)\frac{p_{\alpha'}}{p_\alpha} \left(\frac{m_\alpha}{2\pi\hbar^2}\right)^2 |\tilde{V}(q)|^2$$

$$\times \left| \int_0^\infty u_n^*(r) j_l(qr) u_0(r) r^2 dr \right|^2$$

$$= (2l+1) \frac{p_{\alpha'}}{p_\alpha} \sigma^B(q) \left| \int_0^\infty u_n^*(r) j_l(qr) u_0(r) r^2 dr \right|^2$$

Q18.5 Show that for

$$V = V_0 \quad r \le a$$
$$= 0 \quad r > a$$

the Born approximation (B.A.) is valid if

$$V_0 \ll \frac{p\hbar}{\mu a}, \quad \text{when } \frac{pa}{\hbar} \gg 1$$

and

$$V_0 \ll \frac{\hbar^2}{\mu a^2}, \quad \text{when } \frac{pa}{\hbar} \ll 1.$$

Show further that the first condition is satisfied if

$$\mu \frac{V_0 a^2}{\hbar^2} \ll 1$$

which is also the second condition. Thus B.A. is justified at all energies if $V_0 \ll \frac{\hbar^2}{\mu a^2}$.

Hint: In the first approximation

$$\psi_p^+(\mathbf{r}) = \frac{1}{(2\pi\hbar)^{3/2}} e^{i\mathbf{p}\cdot\mathbf{r}/\hbar}.$$

In the second approximation

$$\psi_p^+(\mathbf{r}) = \frac{1}{(2\pi\hbar)^{3/2}}$$

$$\times \left\{ e^{i\mathbf{p}\cdot\mathbf{r}/\hbar} - \frac{1}{4\pi} \frac{2\mu}{\hbar^2} \int \frac{e^{i\frac{p}{\hbar}|\mathbf{r}-\mathbf{r}'|}}{|\mathbf{r}-\mathbf{r}'|} V(r') e^{i\mathbf{p}\cdot\mathbf{r}'/\hbar} d^3r' \right\}.$$

For B.A. to be valid, the modulus square of the 2nd term in { } should be $\ll 1$. Then calculate { } for the potential given for $\mathbf{r} = 0$. Put $\mathbf{R} = \mathbf{r}' - \mathbf{r}$ and take $\mathbf{p}$ along the z-axis.

Solution: We first calculate the integral in { }

$$\int \frac{e^{i\frac{p}{\hbar}|R|}}{|R|} V(R) e^{i\frac{\mathbf{p}\cdot(\mathbf{R}+\mathbf{r})}{\hbar}} d^3R$$

$$d^3R = 2\pi R^2 dR \sin\theta d\theta$$

$\mathbf{r} = 0$, $\mathbf{p}\cdot\mathbf{R} = pR\cos\theta$, then the integral becomes for the given potential

$$2\pi V_0 \left[\int \frac{e^{i\frac{p}{\hbar}R}}{R} R^2 dR \int_0^\pi e^{i\frac{p}{\hbar}R\cos\theta} \sin\theta d\theta \right]$$

$$= 2\pi V_0 \int_0^a e^{i\frac{p}{\hbar}R} \left[\frac{e^{i\frac{p}{\hbar}R} - e^{-i\frac{p}{\hbar}R}}{i\frac{p}{\hbar}R} \right] R\,dR$$

$$= \frac{2\pi V_0}{i} \frac{\hbar}{p} 2 \int_0^a (e^{2i\frac{p}{\hbar}R} - 1) dR$$

$$= \frac{4\pi V_0}{i} \frac{\hbar}{p} \left[\frac{a}{2i\frac{p}{\hbar}a} \left(e^{2i\frac{p}{\hbar}a} - 1\right) - a \right]$$

Then the second term in { } is

$$-\frac{1}{4\pi} \frac{2\pi V_0}{i} \frac{2\mu}{\hbar^2} \frac{\hbar}{ip} [(-a)], \quad \text{if } \frac{pa}{\hbar} \gg 1$$

$$-\frac{2\pi}{4\pi} \frac{V_0}{i} \frac{2\mu a}{\hbar p} \left[\frac{1}{2!} \left(2i\frac{pa}{\hbar}\right) \right], \quad \text{if } \frac{pa}{\hbar} \ll 1$$

where we have expanded the exponential.
Thus for $\frac{pa}{\hbar} \gg 1$ $B.A$ is valid when

$$V_0 \mu \frac{a}{p\hbar} \ll 1$$

or

$$V_0 \ll \frac{p\hbar}{\mu a}$$

For $\frac{pa}{\hbar} \ll 1$ $B.A$ is valid when

$$\frac{V_0 \mu a}{\hbar p} \left(\frac{p}{\hbar}\right) a \ll 1$$

or

$$V_0 \ll \frac{\hbar^2}{\mu a^2}$$

Now

$$\frac{\mu V_0 a^2}{\hbar^2} \ll 1,$$

which is the second condition. This implies for the first case

$$\frac{\mu V_0 a^2}{\hbar^2} \ll 1 \ll \frac{p_a}{\hbar}$$

or

$$V_0 \ll \frac{p\hbar}{\mu a}$$

which is the first condition. Thus $B.A$ is justified for all energies if $V_0 \ll \frac{\hbar^2}{\mu a^2}$.

Q18.6 Starting from the equation

$$G^{(+)}(E) = \frac{1}{E - H + i\varepsilon},$$

show that

(i)

$$G^+(\mathbf{r}, \mathbf{r}', E) = \langle \mathbf{r} | G^+(E) | \mathbf{r}' \rangle = \sum_n \frac{u_n(\mathbf{r}) u_n^*(\mathbf{r}')}{E - E_n + i\varepsilon}$$

(ii)

$$G^+(t) = \frac{1}{2\pi} \int e^{-iEt/\hbar} G^+(E) dE = -i\theta(t) K(t),$$

where

$$K(t) = \sum_n e^{-iE_n t/\hbar} |E_n\rangle\langle E_n|$$

is called the propagator.

(iii)

$$K(E) = \int e^{iEt/\hbar} K(t) dt = 2\pi \sum_n |E_n\rangle\langle E_n| \delta(E - E_n),$$

where $|E_n\rangle$ are eigenstates of H, $H|E_n\rangle = E_n|E_n\rangle$.

(iv)

$$K(\mathbf{r}, \mathbf{r}', t) = \sum_n e^{-iE_n t/\hbar} u_n(\mathbf{r}) u_n^*(\mathbf{r}')$$

$$K(\mathbf{r}, \mathbf{r}', E) = 2\pi \sum_n u_n(\mathbf{r}) u_n^*(\mathbf{r}') \delta(E - E_n).$$

(v) For a free particle

$$K_0(\mathbf{r}, \mathbf{r}', t) = \left(\frac{m}{2\pi i \hbar t}\right)^{3/2} e^{-\frac{m}{2i\hbar t}(\mathbf{r} - \mathbf{r}')^2}$$

$$K_0(0, 0, E) = \frac{m\sqrt{2mE}}{\pi\hbar^3}.$$

Hint:

$$\sum_n |E_n\rangle\langle E_n| = 1$$

$$\theta(t) = \frac{1}{2\pi i} \int_{-\infty}^{\infty} \frac{e^{i\omega t}}{\omega - i\varepsilon} d\omega.$$

For a free particle

$$u_n(\mathbf{r}) = u_p(\mathbf{r}) = \frac{1}{(2\pi\hbar)^{3/2}} e^{i\mathbf{p}\cdot\mathbf{r}/\hbar}.$$

Solution:

$$G^{(+)}(E) = \frac{1}{E - H + i\epsilon}$$

$$= \sum_n \frac{1}{E - H + i\epsilon} |E_n\rangle\langle E_n|$$

$$= \sum_n \frac{|E_n\rangle\langle E_n|}{E - E_n + i\epsilon}$$

$$G^+(\mathbf{r}, \mathbf{r}', E) = \langle \mathbf{r}|G^+(E)|\mathbf{r}'\rangle$$

$$= \sum_n \frac{\langle \mathbf{r}|E_n\rangle\langle E_n|\mathbf{r}'\rangle}{E - E_n + i\epsilon}$$

$$= \sum_n \frac{u_n(\mathbf{r}) u_n^*(\mathbf{r}')}{E - E_n + i\epsilon}$$

Now

$$G^+(t) = \frac{1}{2\pi} \int e^{-iEt/\hbar} G^+(E) dE$$

$$= \frac{1}{2\pi}(-) \int e^{-iEt/\hbar} \left(\sum_n \frac{|E_n\rangle\langle E_n|}{E_n - E - i\epsilon} \right) dE$$

$$= \frac{-1}{2\pi} \sum_n e^{-iE_n t/\hbar} |E_n\rangle\langle E_n| \int \frac{e^{i(E_n - E)t/\hbar}}{E_n - E - i\epsilon} dE$$

$$= -\frac{1}{2\pi}(2\pi i) \sum_n e^{-iE_n t/\hbar} |E_n\rangle\langle E_n| \theta(t)$$

$$= -i\theta(t)K(t)$$

where

$$K(t) = \sum_n e^{-iE_n t/\hbar} |E_n\rangle\langle E_n| = e^{-iHt/\hbar}$$

$$K(\mathbf{r}, \mathbf{r}', t) = \langle \mathbf{r}|K(t)|\mathbf{r}'\rangle$$

$$= \sum_n e^{-iE_n t/\hbar} u_n(\mathbf{r}) u_n^*(\mathbf{r}')$$

Now

$$K(E) = \int e^{iEt/\hbar} K(t) dt = \sum_n |E_n\rangle\langle E_n| \int e^{i(E-E_n)t/\hbar} dE$$

$$= 2\pi \sum_n |E_n\rangle\langle E_n| \delta(E - E_n)$$

Thus

$$K(\mathbf{r}, \mathbf{r}', E) = 2\pi \sum_n \langle \mathbf{r}|E_n\rangle\langle E_n|\mathbf{r}\rangle \delta(E - E_n)$$

$$= 2\pi \sum_n u_n(\mathbf{r}) u_n^*(\mathbf{r}') \delta(E - E_n)$$

For a free particle

$$u_n(\mathbf{r}) = u_{\mathbf{p}}(\mathbf{r}) = \frac{1}{(2\pi\hbar)^{3/2}} e^{i\mathbf{p}\cdot\mathbf{r}/\hbar},$$

$$E_n \to E_p = \frac{p^2}{2m} \qquad \sum_n \to \int d^3p$$

$$K_0(\mathbf{r}, \mathbf{r}', t) = \frac{1}{(2\pi\hbar)^3} \int d^3p\, e^{\frac{-ip^2 t}{2m\hbar}} e^{i\mathbf{p}\cdot\mathbf{r}/\hbar} e^{-i\mathbf{p}\cdot\mathbf{r}'/\hbar}$$

$$. = \frac{1}{(2\pi\hbar)^3} \int e^{i\mathbf{p}\cdot(\mathbf{r}-\mathbf{r}')/\hbar} e^{-i\frac{p^2 t}{2m\hbar}} d^3p$$

$$p^2 = p_x^2 + p_y^2 + p_z^2$$

$$d^3p = dp_x dp_y dp_z$$

$$I = \int e^{ip_x(x-x')/\hbar} e^{-\frac{ip_x^2 t}{2m\hbar}} dp_x$$

$$= e^{\frac{im}{2\hbar t}(x-x')^2} \int e^{-\frac{it}{2m\hbar}P^2} dP$$

where $P = p_x + \frac{m}{t}(x - x')$ Thus

$$I = e^{-\frac{m}{2i\hbar t}(x-x')^2} \sqrt{\pi \frac{2m\hbar}{ct}}$$

Similarly for p_y, p_z integration. Thus

$$K_0(\mathbf{r}, \mathbf{r}', t) = \frac{1}{(2\pi\hbar)^3} \left(\pi \frac{2m\hbar}{it}\right)^{3/2} e^{-\frac{m}{2i\hbar t}(\mathbf{r}-\mathbf{r}')^2}$$

$$= \left(\frac{m}{2\pi i\hbar t}\right)^{3/2} e^{-\frac{m}{2i\hbar t}(\mathbf{r}-\mathbf{r}')^2}$$

and

$$K_0(\mathbf{r}, \mathbf{r}', E) = 2\pi \frac{1}{(2\pi\hbar)^3} \int d^3p'\, e^{i\mathbf{p}'\cdot(\mathbf{r}-\mathbf{r}')/\hbar} \delta(E - E')$$

where $E' = \frac{p'^2}{2m}$

$$K_0(0, 0, E) = 2\pi \frac{1}{(2\pi\hbar)^3} \int d^3p' \delta(E - E')$$

$$= 2\pi \frac{1}{(2\pi\hbar)^3} \int p'^2 dp' d\Omega \delta(E - E')$$

$$= \frac{2\pi(4\pi)}{(2\pi\hbar)^3} m \int \sqrt{2mE'} \delta(E - E') dE'$$

$$= \frac{1}{\pi\hbar^3} m\sqrt{2mE} \quad .$$

Chapter 19

S-Matrix and Invariance Principles

Q19.1 Show that a state with spin 0 and negative parity cannot decay into 2 spinless particles each having negative parity, if parity conservation is assumed (rotation invariance is of course assumed).

Solution:

$$A \to P_1 + P_2$$

In the rest frame of A

$$\hat{P}|A(0)\rangle = -|A(0)\rangle$$
$$\hat{P}|P_i(\mathbf{p}_i)\rangle = -|P_i(-\mathbf{p}_i)\rangle, \quad i = 1, 2$$

Invariance under space reflection implies

$$\hat{P}T\hat{P}^\dagger = T$$

Now

$$
\begin{aligned}
M(\mathbf{p}_1, \mathbf{p}_2) &= \langle P_1(\mathbf{p}_1), P_2(\mathbf{p}_2)|T|A(0)\rangle \\
&= \langle P_1(\mathbf{p}_1)P_2(\mathbf{p}_2)|\hat{P}^\dagger\hat{P}T\hat{P}^\dagger\hat{P}|A(0)\rangle \\
&= (-1)^2(-1)\langle P_1(-\mathbf{p})P_2(-\mathbf{p}_2)|\hat{P}T\hat{P}^\dagger|A(0)\rangle
\end{aligned}
$$

Hence invariance under space reflection give

$$
\begin{aligned}
M(\mathbf{p}_1, \mathbf{p}_2) &= (-1)^3\langle P_1(-\mathbf{p}_1)P_2(-\mathbf{p}_2)|T|A(0)\rangle \\
&= (-1)^3 M(-\mathbf{p}_1, -\mathbf{p}_2)
\end{aligned}
$$

Because of rotational invariance M can be function of the scalar $\mathbf{p}_1 \cdot \mathbf{p}_2$ which remains unchanged under space reflection i.e.

$$M(\mathbf{p}_1, \mathbf{p}_2) = -M(\mathbf{p}_1, \mathbf{p}_2)$$

or

$$M = 0$$

Thus the decay $A \to P_1 P_2$ is not allowed by parity conservation i.e. it is a parity violating decay.

2nd method:

For initial state, spin of $A = 0$ and parity of $A = -1$. Thus $J_i = 0$, $P_i = -1$

Rotational invariance implies conservation of angular momentum (A.M): $J_f = 0 = L : L$ is the relative orbital A.M of the two particle final state.

Now parity of final state

$$P_f = (-1)(-1)(-1)^{L=0} = +1$$

Hence the decay

$$A \to P_1 P_2$$

is forbidden under parity conservation.

Q19.2 A spinless state decays into two spinless particles, show that (assuming rotational invariance) the matrix elements must be real if time reversal invariance is used.

Solution:

$$\begin{aligned}
M(\mathbf{p}_1, \mathbf{p}_2) &= \langle \mathbf{p}_1, \mathbf{p}_2 | T | A \rangle \\
&= \langle \mathbf{p}_1, \mathbf{p}_2 | \Pi^{-1} \Pi T \Pi^{-1} \Pi | A \rangle \\
&= \langle -\mathbf{p}_1, -\mathbf{p}_2 | T^\dagger | A \rangle^* \\
&= \langle A | T | -\mathbf{p}_1, -\mathbf{p}_2 \rangle
\end{aligned}$$

where we have used definition (i) of time reversal. Now rotational invariance implies that M depends only on $\mathbf{p}_1 \cdot \mathbf{p}_2$. Thus

$$M(\mathbf{p}_1, \mathbf{p}_2) = M^*(\mathbf{p}_1, \mathbf{p}_2)$$

Q19.3 Using either the first or the second definition of time reversal transformation, show that the Schrödinger equation is invariant under time reversal.

Solution: Under time reversal

$$|\psi(t)\rangle \rightarrow |\psi^{(t)}(t)\rangle$$

Thus

$$|\psi^{(t)}\rangle = T|\psi(t)\rangle = \langle \psi(-t)|$$

$$\therefore \langle \mathbf{x}|\psi^{(t)}(t)\rangle = \langle \psi(-t)|\mathbf{x}\rangle = \langle \mathbf{x}|\psi(-t)\rangle^*$$

i.e.

$$\psi^{(t)}(\mathbf{x}, t) = \psi^*(\mathbf{x}, -t)$$

or

$$\psi^{(t)}(\mathbf{x}, -t) = \psi^*(\mathbf{x}, t)$$

The Schrödinger equation is

$$i\hbar \frac{\partial}{\partial t}\psi(\mathbf{x}, t) = H\psi(\mathbf{x}, t)$$

Taking complex conjugate

$$-i\hbar \frac{\partial}{\partial t}\psi^*(\mathbf{x}, t) = H\psi^*(\mathbf{x}, t)$$

Changing $t \rightarrow -t$

$$i\hbar \frac{\partial}{\partial t}\psi^*(\mathbf{x}, -t) = H\psi^*(\mathbf{x}, -t)$$

or

$$i\hbar \frac{\partial}{\partial t}\psi^{(t)}(\mathbf{x}, t) = H\psi^{(t)}(\mathbf{x}, t)$$

Hence the Schrödinger equation is invariant under time reversal.

Chapter 20

Relativistic Quantum Mechanics: Dirac Equation

Q20.1 In the presence of electromagnetic field, Dirac Hamiltonian is given by

$$H = c\boldsymbol{\alpha} \cdot \left(\hat{p} - \frac{e}{c}\mathbf{A}\right) + eA_0 + \beta mc^2.$$

Write

$$\hat{\boldsymbol{\pi}} = \left(\hat{\mathbf{p}} - \frac{e}{c}\mathbf{A}\right).$$

Show that

$$\frac{d\hat{\boldsymbol{\pi}}}{dt} = e(\mathbf{E} + \boldsymbol{\alpha} \times \mathbf{B})$$

where $\mathbf{B} = \boldsymbol{\nabla} \times \mathbf{A}$ is the magnetic field ($\mathbf{A}$ does not depend explicitly on time) and $\mathbf{E}$ is the electric field.

Solution: Heisenberg equation of motion

$$i\hbar\frac{d\hat{\boldsymbol{\pi}}^i}{dt} = [\hat{\boldsymbol{\pi}}^i, H]$$

Note that $\hat{\boldsymbol{\pi}}^i$ commutes with βmc^2. Then

$$\therefore i\hbar\frac{d\hat{\boldsymbol{\pi}}^i}{dt} = e\alpha^j\left(-\frac{e}{c}\right)\{[\hat{p}^i, A^j] + [A^i, p^j]\} + e[\hat{p}^i, A^0]$$

$$= \frac{ie}{\hbar}\alpha^j\left\{\frac{\partial A^j}{\partial x^i} - \frac{\partial A^i}{\partial x^j}\right\} - ie\hbar\frac{\partial A^0}{\partial x^i}$$

$$= -ie\hbar \alpha^j \epsilon^{ijk} (\nabla \times \mathbf{A})^k + ie\hbar E^i$$
$$= ie\hbar[(\boldsymbol{\alpha} \times \mathbf{B})^i + E^i] \tag{20.1}$$

since

$$\mathbf{B} = \nabla \times \mathbf{A}, \quad \mathbf{E} = -\nabla A^0$$

Hence

$$\frac{d\hat{\pi}}{dt} = e(\mathbf{E} + \boldsymbol{\alpha} \times \mathbf{B})$$

Q20.2 From the fundamental properties of γ matrices, derive the following relations:

$$\gamma^\mu (\gamma \cdot a) \gamma_\mu = -2\gamma \cdot a$$
$$\gamma^\mu (\gamma \cdot a)(\gamma \cdot b) \gamma_\mu = 4a \cdot b,$$

where summation over the μ-index is implied and where

$$\gamma \cdot a = \gamma_\nu a^\nu, \ a \cdot b = a_\nu b^\nu;$$

the summation convention is being used.

Solution: (i)

$$\gamma^\mu (\gamma \cdot a) \gamma_\mu = (\gamma^\mu \gamma^\nu \gamma_\mu) a_\nu$$
$$= (2g^{\mu\nu} - \gamma^\nu \gamma^\mu) \gamma_\mu a_\nu$$
$$= 2\gamma^\nu a_\nu - 4\gamma^\nu a_\nu = -2\gamma \cdot a$$

since $\gamma^\mu \gamma_\mu = 4$

(ii)

$$\gamma^\mu (\gamma \cdot a)(\gamma \cdot b) \gamma_\mu = (\gamma^\mu \gamma^\nu \gamma^\lambda) \gamma_\mu a_\nu b_\lambda$$

Now

$$(\gamma^\mu \gamma^\nu \gamma^\lambda) \gamma_\mu = (2g^{\mu\nu} - \gamma^\nu \gamma^\mu) \gamma^\lambda \gamma_\mu$$
$$= 2\gamma^\lambda \gamma^\nu - \gamma^\nu \gamma^\mu \gamma^\lambda \gamma_\mu$$
$$= 2\gamma^\lambda \gamma^\nu - \gamma^\nu (2g^{\mu\lambda} - \gamma^\lambda \gamma^\mu) \gamma_\mu$$
$$= 2(\gamma^\lambda \gamma^\nu + \gamma^\nu \gamma^\lambda) = 4g^{\lambda\nu}$$

Hence

$$\gamma^\mu(\gamma \cdot a)(\gamma \cdot b)\gamma_\mu = 4g^{\lambda\nu}a_\nu b_\lambda = 4a \cdot b$$

Q20.3 Using the anti-commutation relations for γ-matrices, show that

$$(i) \qquad \gamma^5 = -\frac{i}{4!}\varepsilon_{\mu\nu\lambda\rho}\gamma^\mu\gamma^\nu\gamma^\lambda\gamma^\rho$$

$$(ii) \qquad \gamma^5\gamma^\lambda = \frac{i}{3!}\varepsilon^{\mu\nu\rho\lambda}\gamma_\mu\gamma_\nu\gamma_\rho.$$

$$(iii) \qquad [\gamma^\mu, \sigma^{\nu\lambda}] = 2i(g^{\mu\nu}\gamma^\lambda - g^{\mu\lambda}\gamma^\nu)$$

$$(iv) \qquad \{\gamma^\mu, \sigma^{\nu\lambda}\} = -2\varepsilon^{\mu\nu\lambda\rho}\gamma_5\gamma_\rho$$

Solution: We give here some useful identities:

$$\epsilon^{\alpha\beta\gamma\delta}\epsilon_{\alpha\beta\gamma\delta} = -24$$

$$\epsilon^{\alpha\beta\gamma\mu}\epsilon_{\alpha\beta\gamma\nu} = -6\delta^\mu_\nu$$

$$\epsilon^{\alpha\beta\mu\nu}\epsilon_{\alpha\beta\rho\sigma} = -2(\delta^\mu_\rho\delta^\nu_\sigma - \delta^\mu_\sigma\delta^\nu_\rho)$$

where we have used $\epsilon^{123} = 1$ and $= 0$ if any of two indices are equal.

(i)

$$\gamma^5 = -\frac{i}{4!}\epsilon_{\mu\nu\lambda\rho}\gamma^\mu\gamma^\nu\gamma^\lambda\gamma^\rho$$

$$= -\frac{i}{4!}\left[\epsilon_{0jkn}\gamma^0\gamma^j\gamma^k\gamma^n + \epsilon_{i0jk}\gamma^i\gamma^0\gamma^j\gamma^k \right.$$

$$\left. + \epsilon_{ijk0}\gamma^i\gamma^j\gamma^0\gamma^k + \epsilon_{ijk0}\gamma^i\gamma^j\gamma^k\gamma^0\right]$$

$$= -\frac{i}{4!}4\epsilon_{ijk}\gamma^0\gamma^i\gamma^j\gamma^k = \frac{i}{3!}\gamma^0\epsilon^{ijk}\gamma^i\gamma^j\gamma^k$$

$$= \frac{i}{3!}3!(\gamma^0\gamma^1\gamma^2\gamma^3)$$

$$= i\gamma^0\gamma^1\gamma^2\gamma^3$$

(ii) From the identity

$$\gamma^\mu\gamma^\nu\gamma^\lambda = g^{\mu\nu}\gamma^\lambda - g^{\mu\lambda}\gamma^\nu + g^{\nu\lambda}\gamma^\mu + i\epsilon^{\mu\nu\lambda\rho}\gamma_5\gamma_\rho$$

$$(20.2)$$

we derive the results (ii), (iii) and (iv).

Now

$$\epsilon_{\mu\nu\lambda\sigma}\gamma^{\mu}\gamma^{\nu}\gamma^{\lambda} = i\epsilon_{\mu\nu\lambda\sigma}\epsilon^{\mu\nu\lambda\rho}\gamma_5\gamma_\rho$$
$$= i(-6)\delta^{\rho}_{\sigma}\gamma_5\gamma_\rho$$
$$= -3!i\gamma_5\gamma_\sigma$$
$$\gamma_5\gamma_\sigma = \frac{i}{3!}\epsilon_{\mu\nu\lambda\sigma}\gamma^{\mu}\gamma^{\nu}\gamma^{\lambda}$$
$$= \frac{i}{3!}\epsilon^{\mu\nu\lambda\sigma}\gamma_\mu\gamma_\nu\gamma_\lambda = \gamma^5\gamma^\sigma$$

(iii) Now using

$$\gamma^{\nu}\gamma^{\lambda} = g^{\nu\lambda} - i\sigma^{\nu\lambda}$$

we have

$$\gamma^{\mu}\gamma^{\nu}\gamma^{\lambda} = -i\gamma^{\mu}\sigma^{\nu\lambda} + g^{\nu\lambda}\gamma^{\mu}$$
$$\gamma^{\nu}\gamma^{\lambda}\gamma^{\mu} = -i\sigma^{\nu\lambda}\gamma^{\mu} + g^{\nu\lambda}\gamma^{\mu}$$

Thus

$$i[\gamma^{\mu}, \sigma^{\nu\lambda}] = -(\gamma^{\mu}\gamma^{\nu}\gamma^{\lambda} - \gamma^{\nu}\gamma^{\lambda}\gamma^{\mu})$$
$$i\{\gamma^{\mu}, \sigma^{\nu\lambda}\} = 2g^{\nu\lambda}\gamma^{\mu} - (\gamma^{\mu}\gamma^{\nu}\gamma^{\lambda} + \gamma^{\nu}\gamma^{\lambda}\gamma^{\mu})$$

Using (20.2) and noting that $\epsilon^{\nu\lambda\mu\rho} = \epsilon^{\mu\nu\lambda\rho}$, we get

$$[\gamma^{\mu}, \sigma^{\nu\lambda}] = 2i(g^{\mu\nu}\gamma^{\lambda} - g^{\mu\lambda}\gamma^{\nu})$$
$$\{\gamma^{\mu}, \sigma^{\nu\lambda}\} = -2\epsilon^{\mu\nu\lambda\rho}\gamma_5\gamma_\rho$$

Q20.4 Show that in the rest frame of spin $\frac{1}{2}$ particle, the vector

$$n_\mu = \frac{1}{2mc}\varepsilon_{\mu\nu\rho\lambda}\sigma^{\nu\rho}p^\lambda$$

has the value $n_i = \sigma_i$ i.e. $\mathbf{n} = \boldsymbol{\sigma}^i$.

Solution:

$$n_\mu = \frac{1}{2mc}\epsilon_{\mu\nu\rho\lambda}\sigma^{\nu\rho}p^\lambda \qquad (20.3)$$

In the rest frame

$$p^\lambda = (p^0, 0, 0, 0) = (mc, 0, 0, 0)$$

Thus

$$n_\mu = \frac{1}{2mc}\epsilon_{\mu\nu\rho 0}\sigma^{\nu\rho}p^0 = -\frac{1}{2}\epsilon_{0\mu\nu\rho}\sigma^{\nu\rho}$$

$n_0 = 0$ and $n_i = \frac{1}{2}\epsilon_{ijk}\sigma^{jk}$.
Hence

$$n_i = \frac{1}{2}\epsilon_{ijk}\epsilon^{jkl}\sigma^l$$

$$= \frac{1}{2}\epsilon_{ijk}\epsilon_{jkl}\sigma^l$$

$$= \frac{1}{2}(\delta_{kk}\delta_{il} - \delta_{kl}\delta_{ik})\sigma_l$$

$$= \sigma_i \tag{20.4}$$

Q20.5 The Dirac Equation is Lorentz invariant provided that ψ satisfies the transformation law $\psi'(x') = S\psi(x)$, where for the infinitesimal Lorentz transformation

$$S = 1 + \frac{1}{4}\varepsilon_{\mu\nu}\gamma^\mu\gamma^\nu.$$

Consider a 'Pure Lorentz Transformation' in which the primed system is moving with velocity v in the direction x^3-axis and that x'^1, x'^2 are parallel to x^1, x^2

$$x'^1 = x^1$$
$$x'^2 = x^2$$
$$x'^0 = x^0\cosh\omega - x^3\sinh\omega$$
$$x'^3 = x^0\sinh\omega + x^3\cosh\omega,$$

where

$$\cosh\omega = (1 - \beta^2)^{-1/2}$$
$$\sinh\omega = \frac{\beta}{(1 - \beta^2)^{1/2}}, \qquad \beta = \frac{v}{c}.$$

Take ω to be infinitesimal and find S for the above transformation. From this show that for finite ω, S can be written as

$$S = \cosh \omega/2 - i\gamma^3\gamma^0 \sinh \omega/2.$$

Choose the primed system in such a way that it coincides with the rest frame of the electron so that in the primed frame

$$\psi'(x') = \begin{pmatrix} 1 \\ 0 \\ 0 \\ 0 \end{pmatrix} e^{-imc^2t/\hbar}.$$

The above function for the same physical situation in the unprimed frame is given by

$$\psi(x) = S^{-1}\psi'(x').$$

Show that

$$S^{-1}\begin{pmatrix} 1 \\ 0 \\ 0 \\ 0 \end{pmatrix} = \sqrt{\frac{E + mc^2}{2mc^2}} \begin{pmatrix} 1 \\ 0 \\ \frac{p^3 c}{E + mc^2} \\ 0 \end{pmatrix}.$$

Solution:

$$S = e^{-\frac{i}{4}\epsilon_{\mu\nu}\sigma^{\mu\nu}}$$

For pure Lorentz transformation

$$\epsilon_{0i} = \omega_i \qquad \epsilon_{ij} = 0$$

∴ for pure Lorentz transformation

$$\epsilon_{\mu\nu}\sigma^{\mu\nu} = \epsilon_{0\nu}\sigma^{0\nu} + \epsilon_{i\nu}\sigma^{i\nu}$$

$$= \epsilon_{0i}\sigma^{0i} + \epsilon_{i0}\sigma^{i0}$$

$$= 2\epsilon_{0i}\sigma^{0i} = 2\omega_i\frac{i}{2}(\gamma^0\gamma^i - \gamma^i\gamma^0) = -2i\omega n_i\gamma^i\gamma^0.$$

$$S = = e^{\frac{1}{2}\omega n^i\gamma^i\gamma^0}$$

$$= 1 + \frac{1}{2}\omega n^i\gamma^i\gamma^0 + \frac{1}{2!}\left(\frac{\omega}{2}\right)^2 (n^i\gamma^i\gamma^0)(n^j\gamma^j\gamma^0)$$

$$+ \frac{1}{3!}\left(\frac{\omega}{2}\right)^3 (n^i\gamma^i\gamma^0)(n^j\gamma^j\gamma^0)(n^k\gamma^k\gamma^0)\cdots$$

$$= 1 + \frac{1}{2}\omega n^i\gamma^i\gamma^0 - \frac{1}{2!}\left(\frac{\omega}{2}\right)^2 (n^i n^j)\gamma^i\gamma^j$$

$$- \frac{1}{3!}\left(\frac{\omega}{2}\right)^3 n^i\gamma^i\gamma^0(n^j n^k)\gamma^j\gamma^k \cdots$$

$$= \left(1 + \frac{1}{2}(\omega/2)^2 + \cdots\right)$$

$$+ n^i\gamma^i\gamma^0\left(\frac{\omega}{2} + \frac{1}{3!}(\omega/2)^3 + \cdots\right)$$

$$= \cosh\frac{\omega}{2} + n^i\gamma^i\gamma^0 \sinh\frac{\omega}{2}$$

where we have used

$$n^i n^j \gamma^i\gamma^j = \frac{1}{2}n^i n^j(\gamma^i\gamma^j + \gamma^j\gamma^i) = \frac{1}{2}n^i n^j 2g^{ij} = -1$$

In Pauli representation of γ matrices

$$\gamma^i\gamma^0 = \begin{pmatrix} 0 & -\sigma^i \\ -\sigma^i & 0 \end{pmatrix} \tag{20.5}$$

Hence

$$S = \begin{pmatrix} \cosh\frac{\omega}{2} & -\boldsymbol{\sigma}\cdot\mathbf{n}\sinh\frac{\omega}{2} \\ -\boldsymbol{\sigma}\cdot\mathbf{n}\sinh\frac{\omega}{2} & \cosh\frac{\omega}{2} \end{pmatrix}$$

and

$$\psi(x) = S^{-1}\psi'(x')$$

Let x' be a rest frame,

$$u(0) = \begin{pmatrix} \chi^{(r)} \\ 0 \end{pmatrix}$$

In the unprimed frame

$$u(\mathbf{p}) = S^{-1}\begin{pmatrix} \chi^{(r)} \\ 0 \end{pmatrix}$$

$$= \begin{pmatrix} \cosh\frac{\omega}{2}\chi^{(r)} \\ \boldsymbol{\sigma}\cdot\mathbf{n}\sinh\frac{\omega}{2}\chi^{(r)} \end{pmatrix}$$

Now

$$\cosh \omega = \frac{1}{\sqrt{1 - \beta^2}} = \frac{E}{mc^2}$$

$$\sinh \omega = \frac{\beta}{\sqrt{1 - \beta^2}}$$

Then

$$\cosh \frac{\omega}{2} = \frac{1}{\sqrt{2}}(1 + \cosh \omega)^{1/2} = \sqrt{\frac{E + mc^2}{2mc^2}}$$

$$\sinh \frac{\omega}{2} = \frac{1}{\sqrt{2}}(\cosh \omega - 1)^{1/2} = \frac{c|\mathbf{p}|}{\sqrt{2mc^2(E + mc^2)}}$$

Thus

$$u(\mathbf{p}) = \sqrt{\frac{E + mc^2}{2mc^2}} \begin{pmatrix} \chi^{(r)} \\ \frac{c|\mathbf{p}|}{E+mc^2} \boldsymbol{\sigma} \cdot \mathbf{n}\chi^{(r)} \end{pmatrix}$$

$$= \sqrt{\frac{E + mc^2}{2mc^2}} \begin{pmatrix} \chi^{(r)} \\ \frac{c\boldsymbol{\sigma} \cdot \mathbf{p}}{E+mc^2} \chi^{(r)} \end{pmatrix}$$

For the Lorentz boost along z-direction

$$\boldsymbol{\sigma} \cdot \mathbf{p} = p^3 \sigma^3 = \begin{pmatrix} p^3 & 0 \\ 0 & -p^3 \end{pmatrix} = p\sigma^3$$

For the case

$$\chi^{(r)} = \begin{pmatrix} 1 \\ 0 \end{pmatrix}$$

$$u(0) = \begin{pmatrix} 1 \\ 0 \\ 0 \\ 0 \end{pmatrix}$$

$$u(\mathbf{p}) = \sqrt{\frac{E + mc^2}{2mc^2}} \begin{pmatrix} 1 \\ 0 \\ \frac{p^3}{E+mc^2} \\ 0 \end{pmatrix}$$

We have solved this problem for a general case, rather than for the special case of problem 20.5.

Q20.6 In Dirac theory of the electron, one identifies the electron intrinsic spin, **S** as

$$\mathbf{S} = \hbar \frac{\boldsymbol{\alpha} \times \boldsymbol{\alpha}}{4i}$$

Using only the anti-commutation rules for Dirac matrices α^1, α^2 and α^3, show that the electron has spin $1/2$ i.e. show $S^2 = \dfrac{3\hbar^2}{4}$

Solution:

$$S^2 = S_i S_i = -\frac{\hbar^2}{4}(\epsilon_{ijk}\alpha_j\alpha_k)(\epsilon_{ij'k'}\alpha_{j'}\alpha_{k'})$$

$$= -\frac{\hbar^2}{16}\epsilon_{ijk}\epsilon_{ij'k'}\alpha_j\alpha_k\alpha_{j'}\alpha_{k'}$$

$$= -\frac{\hbar^2}{16}(\delta_{jj'}\delta_{kk'} - \delta_{jk'}\delta_{kj'})\alpha_j\alpha_k\alpha_{j'}\alpha_{k'}$$

$$= -\frac{\hbar^2}{16}(\alpha_j\alpha_k\alpha_j\alpha_k - \alpha_j\alpha_k\alpha_k\alpha_j)$$

$$= -\frac{\hbar^2}{16}(\alpha_j\alpha_k(\alpha_j\alpha_k + \alpha_k\alpha_j) - 2\alpha_j\alpha_k\alpha_k\alpha_j)$$

$$= -\frac{\hbar^2}{16}(-4\alpha_j\alpha_j)$$

$$= \frac{3\hbar^2}{4}$$

Eigenvalues of S^2 are

$$s(s+1)\hbar^2$$

Now for

$$s = \frac{1}{2}, \qquad s(s+1)\hbar^2 = \frac{3}{4}\hbar^2$$

Hence the Dirac spinor ψ corresponds to spin $\frac{1}{2}$ particle.

Q20.7 The large and small components of the solution of the Dirac equation

$$(i\gamma_\mu \hat{p}_\mu + mc)\psi = \frac{V}{c}\psi$$

are defined as $\psi_A_B = \frac{1}{2}(1 \pm \gamma^0)\psi$.

(We choose a rep. in which $\gamma_0 = \beta$ is diagonal.)

Obtain the equation for ψ_A and show that when the kinetic energy, T, of the particle $(cp_0 = mc^2 + T)$ is such that $T \ll mc^2$, ψ_A reduces to the non-relativistic Schrödinger wave function multiplied by $e^{-imc^2 t/\hbar}$.

Solution:

$$\left(i\hbar\gamma^\mu \partial_\mu - mc + \frac{V}{c}\right)\psi = 0$$

Now from Eqs. (20.270) of the text, with $mc \to mc - \frac{V}{c}$, we have for the stationary state

$$\psi(\mathbf{x}, t) = \psi(\mathbf{x})e^{-iEt/\hbar}$$

$$i\hbar\frac{\partial}{\partial t}\psi(\mathbf{x}, t) = i\hbar(-iE/\hbar)\psi(\mathbf{x}, t)$$

$$= E\psi(\mathbf{x}, t) \tag{20.6}$$

$$\phi_B = -i\hbar c\frac{\boldsymbol{\sigma}\cdot\boldsymbol{\nabla}\phi_A}{E + mc^2 - V} \tag{20.7}$$

$$i\hbar\boldsymbol{\sigma}\cdot\boldsymbol{\nabla}\phi_B = -\left(\frac{E}{c} - mc + \frac{V}{c}\right)\phi_A \tag{20.8}$$

In the non-relativistic limit

$$T \ll mc^2, \qquad V \ll mc^2$$

$$\phi_B \approx \frac{-i\hbar}{2mc}\boldsymbol{\sigma}\cdot\boldsymbol{\nabla}\phi_A \tag{20.9}$$

Substituting (20.9) in (20.8)

$$(E - mc^2 + V)\phi_A + \frac{\hbar^2(\boldsymbol{\sigma}\cdot\boldsymbol{\nabla})(\boldsymbol{\sigma}\cdot\boldsymbol{\nabla})}{2m}\phi_A = 0$$

or

$$-\frac{\hbar}{2m}\nabla^2\phi_A = (T + V)\phi,$$

which is the Schrödinger equation. (We reserve ψ_A for four component spinor and ϕ_A for two component spinor)

Q20.8 Show that, using the Dirac equation,

$$\sum_s \bar{u}^s(p)\gamma^\mu u^s(p) = \frac{2}{m}p^\mu,$$

where $s = 1,2$ is the spin index. Show also that

$$p_\mu(\bar{u}(p)\sigma^{\mu\lambda}u(p)) = 0$$

while

$$p^\lambda(\bar{u}(p)\gamma_5\sigma^{\mu\nu}u(p)) = imc\bar{u}(p)\gamma^\mu\gamma^5 u(p)$$

Solution: Dirac equation

$$(\not{p} - mc)u^s(\mathbf{p}) = 0 \qquad (20.10)$$

$$\bar{u}^s(\mathbf{p})(\not{p} - mc) = 0 \qquad (20.11)$$

Multiply Eq. (20.10) by $\bar{u}(\mathbf{p})\gamma^\nu$ on the left and multiply (20.11) on the right by $\gamma^\nu u(\mathbf{p})$ and add:

$$\bar{u}^s(\mathbf{p})\gamma^\nu(\not{p} - mc)u^s(\mathbf{p}) + \bar{u}^s(\mathbf{p})(\not{p} - mc)\gamma^\nu u^s(\mathbf{p}) = 0$$

or

$$\bar{u}^s(\mathbf{p})(\gamma^\nu\not{p} + \not{p}\gamma^\nu)u^s(\mathbf{p}) = 2mc\bar{u}^s(\mathbf{p})\gamma^\nu u^s(\mathbf{p})$$

Now the L.H.S

$$\bar{u}^s(\mathbf{p})(\gamma^\nu\gamma^\lambda + \gamma^\lambda\gamma^\nu)p_\lambda u^s(\mathbf{p}) = 2g^{\nu\lambda}p_\lambda\bar{u}^s(\mathbf{p})u^s(\mathbf{p})$$
$$= 2p^\nu\bar{u}^s(\mathbf{p})u^s(\mathbf{p})$$

Hence, we have

$$2mc\bar{u}^s(\mathbf{p})\gamma^\mu u^s(\mathbf{p}) = 2p^\mu\bar{u}^s(\mathbf{p})u^s(\mathbf{p})$$

or

$$\sum_s \bar{u}^s(\mathbf{p})\gamma^\mu u^s(\mathbf{p}) = \frac{2p^\mu}{mc}$$

(ii) Now

$$p_\mu(\bar{u}^s(\mathbf{p})\sigma^{\mu\lambda}u^s(\mathbf{p}))$$

$$= p_\mu \frac{i}{2}\bar{u}^s(\mathbf{p})(\gamma^\mu\gamma^\lambda - \gamma^\lambda\gamma^\mu)u^s(\mathbf{p})$$

$$= \frac{i}{2}\bar{u}(\mathbf{p})(\slashed{p}\gamma^\lambda - \gamma^\lambda\slashed{p})u(\mathbf{p})$$

$$= \frac{i}{2}(mc\bar{u}(\mathbf{p})\gamma^\lambda u(\mathbf{p}) - mc\bar{u}(\mathbf{p})\gamma^\lambda u(\mathbf{p})) = 0$$

$$p_\lambda(\bar{u}^s(\mathbf{p})\gamma^5\sigma^{\mu\lambda}u^s(\mathbf{p}))$$

$$= \frac{i}{2}(\bar{u}(\mathbf{p})\gamma^5\gamma^\mu\slashed{p}u(\mathbf{p}) - \bar{u}(\mathbf{p})\gamma^5\slashed{p}\gamma^\mu u(\mathbf{p}))$$

$$= \frac{i}{2}mc[\bar{u}(\mathbf{p})(\gamma^5\gamma^\mu + \gamma^5\gamma^\mu)u(\mathbf{p})]$$

$$= imc\bar{u}(\mathbf{p})\gamma^5\gamma^\mu u(\mathbf{p})$$

Q20.9 A positive energy Dirac spinor (using the Pauli rep. of γ matrices) is given by

$$u^{(r)}(\mathbf{p}) = \begin{pmatrix} \chi^{(r)} \\ \frac{c}{E+mc^2}\boldsymbol{\sigma}\cdot\mathbf{p}\chi^{(r)} \end{pmatrix}.$$

As $m \to 0$, show that the γ_5 operator and the helicity $\frac{\boldsymbol{\sigma}\cdot\mathbf{p}}{|\mathbf{p}|}$ operators have the same effect on $u^{(r)}(\mathbf{p})$.

Solution:

$$\gamma^5 u^{(r)}(\mathbf{p}) = \begin{pmatrix} 0 & 1 \\ 1 & 0 \end{pmatrix}\begin{pmatrix} \chi^{(r)} \\ \frac{c\boldsymbol{\sigma}\cdot\mathbf{p}}{E}\chi^{(r)} \end{pmatrix}$$

$$= \begin{pmatrix} \frac{\boldsymbol{\sigma}\cdot\mathbf{p}}{|\mathbf{p}|}\chi^{(r)} \\ \chi^{(r)} \end{pmatrix}$$

Helicity operator $\mathcal{H} = \frac{\boldsymbol{\sigma}\cdot\mathbf{p}}{|\mathbf{p}|}$

$$\mathcal{H}u^{(r)}(\mathbf{p}) = \begin{pmatrix} \frac{\boldsymbol{\sigma}\cdot\mathbf{p}}{|\mathbf{p}|} & 0 \\ 0 & \frac{\boldsymbol{\sigma}\cdot\mathbf{p}}{|\mathbf{p}|} \end{pmatrix}\begin{pmatrix} \chi^{(r)} \\ \frac{\boldsymbol{\sigma}\cdot\mathbf{p}}{|\mathbf{p}|}\chi^{(r)} \end{pmatrix}$$

$$= \begin{pmatrix} \frac{\boldsymbol{\sigma}\cdot\mathbf{p}}{|\mathbf{p}|}\chi^{(r)} \\ \chi^{(r)} \end{pmatrix} \tag{20.12}$$

Q20.10 In the approximation $E \approx mc^2$, $|eA_0| \ll mc^2$, show the equation for ψ_A is given by

$$\left[\frac{1}{2m} (\mathbf{p} - \frac{e}{c}\mathbf{A})^2 - \frac{e\hbar}{2mc}\boldsymbol{\sigma} \cdot \mathbf{B} + eA_0 \right] \psi_A = E^{(NR)}\psi_A,$$

where $E^{(NR)} = E - mc^2$ and $\mathbf{B} = \boldsymbol{\nabla} \times \mathbf{A}$. Take $\mathbf{A} = 0 = A_0$. Consider the probability current 4 vector

$$j^\mu = e\bar{\psi}\gamma^\mu\psi.$$

Show that in the lowest order $(E \approx mc^2)$

$$j^0 = c\psi_A^\dagger\psi_A$$
$$j^i = c\psi^\dagger\alpha^i\psi$$
$$= -\frac{i\hbar}{2m}\left\{ \psi_A^\dagger\frac{\partial\psi_A}{\partial x^i} - \frac{\partial\psi_A^\dagger}{\partial x^i}\psi_A \right\}$$

plus a term $i\varepsilon^{ijk}\frac{\partial}{\partial x^j}(\psi_A^\dagger\sigma_k\psi_A)$ which can be ignored since when j^i is integrated over d^3x, it gives zero.

Solution: We start from Eq. (20.168) of the text with

$$i\hbar D^\mu = (p^\mu - \frac{e}{c}A^\mu)$$

which gives

$$\left[\left(\hat{p}^\mu - \frac{e}{c}A^\mu \right) \left(\hat{p}_\mu - \frac{e}{c}A_\mu \right) + \frac{e\hbar}{c}\boldsymbol{\Sigma} \cdot \mathbf{B} \right.$$
$$\left. - \frac{ie\hbar}{c}\boldsymbol{\alpha} \cdot \mathbf{E} - m^2c^2 \right] \psi = 0 \tag{20.13}$$

Since $\mathbf{E} = \boldsymbol{\nabla}A_0, eA_0 \ll mc^2$, the term $\boldsymbol{\alpha} \cdot \mathbf{E}$ can be neglected. Now

$$\psi = \begin{pmatrix} \phi_A \\ \phi_B \end{pmatrix} \tag{20.14}$$

and replacing $\hat{p}_0 \equiv \frac{\hat{E}}{c}$ by $\frac{E}{c}$ for stationary states, we get from Eq. (20.13)

$$\left[(E - eA^0)^2 - \left(\hat{\mathbf{p}} - \frac{e}{c}\mathbf{A} \right)^2 + \frac{e\hbar}{c}\boldsymbol{\sigma} \cdot \mathbf{B} - m^2c^2 \right] \phi_A = 0$$
$$\tag{20.15}$$

Now in the non relativistic limit

$$(E - eA_0)^2 - m^2c^2$$
$$= (E - eA_0 + mc^2)(E - eA_0 - mc^2)$$
$$\simeq 2mc^2(E^{NR} - eA_0)$$

Hence from Eq. (20.15), we get

$$\left[\frac{1}{2m} \left(\hat{\mathbf{p}} - \frac{e}{c}\mathbf{A} \right)^2 + eA_0 - \frac{e\hbar}{2mc}\boldsymbol{\sigma} \cdot \mathbf{B} \right] \phi_A = E^{NR}\phi_A$$

(20.16)

Thus for the slowly moving particle of mass m and charge e, the Hamiltonian is

$$H = \frac{1}{2m}(\hat{\mathbf{p}} - \frac{e}{c}\mathbf{A})^2 + eA_0 - \frac{e\hbar}{2mc}\boldsymbol{\sigma} \cdot \mathbf{B}$$ (20.17)

This differs from the Schrodinger Hamiltonian in the term $-\frac{e\hbar}{2mc}\boldsymbol{\sigma} \cdot \mathbf{B} = \boldsymbol{\mu} \cdot \mathbf{B}$. Hence this term shows the interaction of a particle of magnetic moment $\boldsymbol{\mu} = \frac{e\hbar}{2mc}\boldsymbol{\sigma}$ with the magnetic field. We conclude that wave function ϕ_A represents a particle with intrinsic spin $\frac{1}{2}$ and intrinsic magnetic moment $\frac{e\hbar}{2mc}\boldsymbol{\sigma}$. Now

$$j^\mu = c\bar{\psi}\gamma^\mu\psi$$
$$j^0 = c\bar{\psi}\gamma^0\psi = c\psi^\dagger\psi$$
$$= c \begin{pmatrix} \phi_A^\dagger & \phi_B^\dagger \end{pmatrix} \begin{pmatrix} \phi_A \\ \phi_B \end{pmatrix}$$
$$= c\phi_A^\dagger\phi_A + c\phi_B^\dagger\phi_B$$
$$= c\phi_A^\dagger\phi_A + \mathcal{O}(v^2/c^2)$$
$$j^i = c\bar{\psi}\gamma^i\psi$$
$$= c\psi^\dagger\gamma^0\gamma^i\psi = c\psi^\dagger\alpha^i\psi$$
$$= c \begin{pmatrix} \phi_A^+ & \phi_B^+ \end{pmatrix} \begin{pmatrix} 0 & \sigma^i \\ \sigma^i & 0 \end{pmatrix} \begin{pmatrix} \phi_A \\ \phi_B \end{pmatrix}$$
$$= c(\phi_A^+\sigma^i\phi_B + \phi_B^+\sigma^i\phi_A)$$

$$= \frac{1}{2m} \phi_A^+ \left(-i\hbar\sigma^i\sigma^j \frac{\partial}{\partial x^j} \right) \phi_A$$

$$+ \frac{1}{2m} \left(\frac{\partial}{\partial x^j} \cdot \phi_A^+ \right) (i\hbar\sigma^i\sigma^j)\phi_A$$

where we have used Eq. (20.271) of the text.
Now

$$\sigma^i\sigma^j = \delta^{ij} + i\epsilon^{ijk}\sigma^k$$

Thus

$$j^i = -\frac{i\hbar}{2m} \left\{ \phi_A^+ \frac{\partial\phi_A}{\partial x^i} - \frac{\partial\phi_A^+}{\partial x^i}\phi_A \right\}$$

plus a term

$$i\epsilon^{ijk} \frac{\partial}{\partial x^i} (\phi_A^+\sigma^k\phi_A)$$

Q20.11 Show that under charge conjugation

$$\phi_L^c = -i\sigma^2\phi_R^*$$
$$\phi_R^c = i\sigma^2\phi_L^*$$

On the other hand under parity

$$\phi_L \leftrightarrow \phi_R$$

Thus under CP

$$\phi_L \to -i\sigma^2\phi_L^*$$
$$\phi_R \to i\sigma^2\phi_R^*$$

Solution: Under the charge conjugation from Eq. (20.255) of the text

$$\psi^c = -i\gamma^2\psi^*$$

Using chiral representatiion of $\gamma-$matrices

$$\begin{pmatrix} \phi_L^c \\ \phi_R^c \end{pmatrix} = -i \begin{pmatrix} 0 & \sigma^2 \\ -\sigma^2 & 0 \end{pmatrix} \begin{pmatrix} \phi_L^* \\ \phi_R^* \end{pmatrix} \qquad (20.18)$$

Hence

$$\phi_L^c = -i\sigma^2 \phi_R^*$$
$$\phi_R^c = i\sigma^2 \phi_L^*$$

Now under parity

$$\psi \rightarrow \gamma^0 \psi$$

$$\begin{pmatrix} \phi_L \\ \phi_R \end{pmatrix} \rightarrow \begin{pmatrix} 0 & 1 \\ 1 & 0 \end{pmatrix} \begin{pmatrix} \phi_L \\ \phi_R \end{pmatrix} = \begin{pmatrix} \phi_R \\ \phi_L \end{pmatrix} \qquad (20.19)$$

Hence under parity

$$\phi_L \leftrightarrow \phi_R$$

Thus under CP

$$\phi_L \rightarrow -i\sigma^2 \phi_L^*$$
$$\phi_R \rightarrow i\sigma^2 \phi_R^*$$

Q20.12 Consider the coupled 2-component equations

$$i\sigma_L^\mu \partial_\mu \phi_L - \frac{mc}{\hbar c} \phi_R = 0$$

$$i\sigma_R^\mu \partial_\mu \phi_R - \frac{mc}{\hbar c} \phi_L = 0$$

where

$$\sigma_L^\mu = (1, -\boldsymbol{\sigma}) = \bar{\sigma}^\mu, \quad \sigma_R^\mu = (1, \boldsymbol{\sigma}) = \sigma^\mu$$

Write the above equation in the 4-component form

$$\left(i\gamma'^\mu \partial_\mu - \frac{mc}{\hbar} \right) \psi' = 0$$

where

$$\psi' = \begin{pmatrix} \phi_L \\ \phi_R \end{pmatrix}$$

Obtain the equivalent form of γ'^μ and $\gamma_5' = i\gamma_0' \gamma_1' \gamma_2' \gamma_3'$ and check that

$$(\gamma'^\mu \gamma'^\nu + \gamma'^\nu \gamma'^\mu) = 2g^{\mu\nu}$$

Verify that the Weyl set $\{\gamma'^\mu\}$ and the Pauli set $\{\gamma^\mu\}$ are related by

$$\gamma'^\mu = S\gamma^\mu S^{-1}$$

where

$$S = \frac{1}{\sqrt{2}}\begin{pmatrix} 1 & -1 \\ 1 & 1 \end{pmatrix}$$

Solution: From the given equtions

$$i\partial_0\phi_L - i\sigma^i\partial_i\phi_L - \frac{mc}{\hbar}\phi_R = 0$$

$$i\partial_0\phi_R + i\sigma^i\partial_i\phi_R - \frac{mc}{\hbar}\phi_L = 0$$

These can be combined

$$i\partial_0\begin{pmatrix}\phi_R \\ \phi_L\end{pmatrix} + i\partial_i\begin{pmatrix}\sigma^i\phi_R \\ -\sigma^i\phi_L\end{pmatrix} - \frac{mc}{\hbar}\begin{pmatrix}\phi_L \\ \phi_R\end{pmatrix} = 0$$

Introduce

$$\gamma^{0\prime} = \begin{pmatrix} 0 & 1 \\ 1 & 0 \end{pmatrix}, \quad \gamma'^i = \begin{pmatrix} 0 & \sigma^i \\ -\sigma^i & 0 \end{pmatrix}$$

Then the above equation can be written as

$$\left[i\partial_0\begin{pmatrix} 0 & 1 \\ 1 & 0 \end{pmatrix} + i\partial_i\begin{pmatrix} 0 & \sigma^i \\ -\sigma^i & 0 \end{pmatrix} - \frac{mc}{\hbar}\right]\begin{pmatrix}\phi_L \\ \phi_R\end{pmatrix} = 0$$

This takes the 4-component form

$$\left(i\gamma'^\mu\partial_\mu - \frac{mc}{\hbar}\right)\psi' = 0$$

where

$$\psi' = \begin{pmatrix}\phi_L \\ \phi_R\end{pmatrix}$$

It is easy to check as has been done in the text that

$$\gamma'^\mu\gamma'^\nu + \gamma'^\nu\gamma'^\mu = 2g^{\mu\nu}$$

$$\gamma'^5 = i\gamma'^0\gamma'^1\gamma'^2\gamma'^3 = \begin{pmatrix} -1 & 0 \\ 0 & 1 \end{pmatrix}$$

$$S = \frac{1}{\sqrt{2}} \begin{pmatrix} 1 & -1 \\ 1 & 1 \end{pmatrix}$$

$$\gamma'^0 = S\gamma^0 S^{-1}$$

$$= \frac{1}{2} \begin{pmatrix} 1 & -1 \\ 1 & 1 \end{pmatrix} \begin{pmatrix} 1 & 0 \\ 0 & -1 \end{pmatrix} \begin{pmatrix} 1 & 1 \\ -1 & 1 \end{pmatrix}$$

$$= \frac{1}{2} \begin{pmatrix} 1 & -1 \\ 1 & 1 \end{pmatrix} \begin{pmatrix} 1 & 1 \\ 1 & -1 \end{pmatrix}$$

$$= \frac{1}{2} \begin{pmatrix} 0 & 2 \\ 2 & 0 \end{pmatrix} = \begin{pmatrix} 0 & 1 \\ 1 & 0 \end{pmatrix}$$

$$\gamma'^i = S \begin{pmatrix} 0 & \sigma^i \\ -\sigma^i & 0 \end{pmatrix} S^{-1}$$

$$= \frac{1}{2} \begin{pmatrix} 1 & -1 \\ 1 & 1 \end{pmatrix} \begin{pmatrix} 0 & \sigma^i \\ -\sigma^i & 0 \end{pmatrix} \begin{pmatrix} 1 & 1 \\ -1 & 1 \end{pmatrix}$$

$$= \frac{1}{2} \begin{pmatrix} 1 & -1 \\ 1 & 1 \end{pmatrix} \begin{pmatrix} -\sigma^i & \sigma^i \\ -\sigma^i & -\sigma^i \end{pmatrix}$$

Thus

$$\gamma'^i = \begin{pmatrix} 0 & \sigma^i \\ -\sigma^i & 0 \end{pmatrix}$$

Additional Problems:

Q1 Show that

(i)

$$Tr(\gamma^\mu \gamma^\nu) = 4g^{\mu\nu}$$

(ii)

$$Tr(\gamma^\mu \gamma^\nu \gamma^\rho \gamma^\sigma) = 4[g^{\sigma\rho} g^{\mu\nu} - g^{\nu\sigma} g^{\mu\rho} + 2g^{\mu\sigma} g^{\nu\rho}]$$

(iii) Trace of odd number of γ-matrices to zero.

Solution: The following properties are useful

$$Tr(AB) = Tr(BA);$$

$$Tr(ABC) = Tr(BAC) = Tr(BCA): \quad \text{cyclic property}$$

γ-matrices are traceless

$$Tr(\gamma^\mu) = 0, \quad \mu = 0, 1, 2, 3$$
$$Tr(\gamma^5) = 0$$

(i) Now $Tr(\gamma^\mu\gamma^\nu) = Tr(\gamma^\nu\gamma^\mu)$ and from

$$\gamma^\mu\gamma^\nu + \gamma^\nu\gamma^\mu = 2g^{\mu\nu}$$
$$Tr(\gamma^\mu\gamma^\nu) = g^{\mu\nu}Tr(I)$$
$$= 4g^{\mu\nu}$$

(ii)

$$\gamma^\mu\gamma^\nu\gamma^\rho\gamma^\sigma$$
$$= \gamma^\mu\gamma^\nu(2g^{\rho\sigma} - \gamma^\sigma\gamma^\rho)$$
$$= 2g^{\rho\sigma}\gamma^\mu\gamma^\nu - \gamma^\mu(2g^{\nu\sigma} - \gamma^\sigma\gamma^\nu)\gamma^\rho$$
$$= 2g^{\rho\sigma}\gamma^\mu\gamma^\nu - 2g^{\nu\sigma}\gamma^\mu\gamma^\rho + (2g^{\mu\sigma} - \gamma^\sigma\gamma^\mu)\gamma^\nu\gamma^\rho$$
$$= 2g^{\rho\sigma}\gamma^\mu\gamma^\nu - 2g^{\nu\sigma}\gamma^\mu\gamma^\rho + 2g^{\mu\sigma}\gamma^\nu\gamma^\rho - \gamma^\sigma\gamma^\mu\gamma^\nu\gamma^\rho$$
$$\therefore \ \gamma^\mu\gamma^\nu\gamma^\rho\gamma^\sigma + \gamma^\sigma\gamma^\mu\gamma^\nu\gamma^\rho$$
$$= 2g^{\rho\sigma}\gamma^\mu\gamma^\nu - 2g^{\nu\sigma}\gamma^\mu\gamma^\rho + 2g^{\mu\sigma}\gamma^\nu\gamma^\rho$$

Now

$$Tr(\gamma^\sigma\gamma^\mu\gamma^\nu\gamma^\rho) = Tr(\gamma^\mu\gamma^\nu\gamma^\rho\gamma^\sigma)$$

Thus

$$Tr(\gamma^\mu\gamma^\nu\gamma^\rho\gamma^\sigma) = 4(g^{\rho\sigma}g^{\mu\nu} - g^{\nu\sigma}g^{\mu\rho} + g^{\mu\sigma}g^{\nu\sigma})$$

(iii) Now

$$Tr(\gamma^\mu\gamma^\nu \cdots \gamma^\lambda) = Tr((\gamma^5)^2\gamma^\mu\gamma^\nu \cdots \gamma^\lambda)$$
$$= (-1)^n Tr(\gamma^5\gamma^\mu\gamma^\nu \cdots \gamma^\lambda\gamma^5)$$
$$= (-1)^n Tr(\gamma^5\gamma^5\gamma^\mu\gamma^\nu \cdots \gamma^\lambda)$$
$$= (-1)^n Tr(\gamma^\mu\gamma^\nu \cdots \gamma^\lambda)$$

Hence for odd n,

$$Tr(\gamma^\mu \gamma^\nu \cdots \gamma^\lambda) = 0$$

Mott-Scattering

Q2 Consider the scattering of electron by a spinless charged particle of charge Ze *viz* the scattering of electrons by a coulomb potential

$$V(r) = \frac{Ze}{r}$$

Show that

$$\frac{d\sigma}{d\Omega} = \frac{4Z^2 e^4}{(cq)^4} E^2 \left(1 - \beta^2 \sin^2 \frac{\theta}{2}\right)$$

where

$$\mathbf{q}^2 = (\mathbf{p}_i - \mathbf{p}_f)^2 = 2|\mathbf{p}|^2(1 - \cos^2 \theta)$$

$$= 4|\mathbf{p}|^2 \sin^2 \frac{\theta}{2}$$

$$\beta = \frac{v}{c} = \frac{c|\mathbf{p}|}{E}; \quad |\mathbf{p}_i| = |\mathbf{p}_f| = |\mathbf{p}| \quad E_i = E_f = E;$$

elastic scattering

Solution: In the Born approximation, the scattering amplitude is given by

$$T_{fi} = -e\langle f|V|i\rangle$$

$$T_{fi} = -e \int \langle f|\mathbf{r}'\rangle \langle \mathbf{r}'|V|\mathbf{r}\rangle \langle \mathbf{r}|i\rangle d^3r d^3r'$$

$$- e \int \psi_f^\dagger(\mathbf{r}) V(\mathbf{r}) \psi_i(\mathbf{r}) d^3r$$

where we have used $\langle \mathbf{r}'|V|\mathbf{r}\rangle = V(r)\delta^3(\mathbf{r} - \mathbf{r}')$

Taking the electron as a Dirac particle

$$\psi(x) = \left(\frac{1}{2\pi\hbar}\right)^{3/2} \sqrt{\frac{mc}{p_0}} e^{ip.x/\hbar} u(\mathbf{p})$$

$$= \left(\frac{1}{2\pi\hbar}\right)^{3/2} \sqrt{\frac{mc}{p_0}} e^{iE.t/c\hbar} e^{-i\mathbf{p}.\mathbf{r}/\hbar} u(\mathbf{p})$$

We consider the elastic scattering

$$E_f = E_i, \quad |\mathbf{p}_f| = |\mathbf{p}_i| = |\mathbf{p}|$$
$$p_{io} = E/c = p_{fo}$$

Thus

$$\psi_f^\dagger(\mathbf{r}) = \left(\frac{1}{2\pi\hbar}\right)^{3/2} \sqrt{\frac{mc}{p_{fo}}} e^{i\mathbf{p}_f \cdot \mathbf{r}/\hbar} u^\dagger(\mathbf{p}_f)$$

$$\psi_i(\mathbf{r}) = \left(\frac{1}{2\pi\hbar}\right)^{3/2} \sqrt{\frac{mc}{p_{io}}} e^{-i\mathbf{p}_i \cdot \mathbf{r}/\hbar} u(\mathbf{p}_i) \qquad (20.20)$$

$$T = -\left(\frac{1}{2\pi\hbar}\right)^3 \frac{mc}{\sqrt{p_{fo}p_{io}}} e u^\dagger(\mathbf{p}_f) u(\mathbf{p}_i)$$

$$\times \int e^{i(\mathbf{p}_f - \mathbf{p}_i)\cdot\mathbf{r}/\hbar} V(\mathbf{r}) d^3 r$$

where

$$\int e^{i(\mathbf{p}_f - \mathbf{p}_i)\cdot\mathbf{r}/\hbar} V(r) d^3 r = \int e^{-i(\mathbf{p}_i - \mathbf{p}_f)\cdot\mathbf{r}/\hbar} \left(-\frac{Ze}{|\mathbf{r}|}\right) d^3 x$$

$$= -Ze \frac{4\pi\hbar^2}{|\mathbf{p}_i - \mathbf{p}_f|^2}$$

Using Eq. (18.66) of the text,

$$d\sigma = \frac{1}{\text{Flux}} \frac{2\pi}{\hbar} \rho_f(E_f) |T_{fi}|^2 \qquad (20.21)$$

where

$$T_{fi} = \left(\frac{1}{2\pi\hbar}\right)^3 \left(\frac{mc^2}{E}\right) (Ze^2) 4\pi\hbar^2 \frac{1}{|\mathbf{p}_i - \mathbf{p}_f|^2} u^\dagger(\mathbf{p}_f) u(\mathbf{p}_i)$$

and

$$\rho_f(E_f) = |\mathbf{p}_f|^2 \frac{d|\mathbf{p}_f|}{dE_f} d\Omega$$

For our case $|\mathbf{p}_i| = |\mathbf{p}_f| = |\mathbf{p}|$

$$E_f = E_i = E, \quad E^2 = c^2 |\mathbf{p}|^2 + m^2 c^4$$

so that

$$\rho_f(E_f) = \frac{|\mathbf{p}|E}{c^2} d\Omega$$

$$(\text{Flux})_{\text{in}} = \frac{1}{(2\pi\hbar)^3} v, \qquad v = \frac{c^2|\mathbf{p}|}{E} \qquad (20.22)$$

Thus

$$\frac{d\sigma}{d\Omega} = \frac{2\pi}{\hbar}(2\pi\hbar)^3 \frac{E^2}{c^4} \frac{Z^2 e^4}{(2\pi\hbar)^6} \frac{m^2 c^4}{E^2} \frac{16\pi^2\hbar^4}{|\mathbf{p}_f - \mathbf{p}_i|^4} |F_{fi}|^2$$

$$= \frac{m^2 4 Z^2 e^4}{(\mathbf{p}_i - \mathbf{p}_f)^4} |F_{fi}|^2 \qquad (20.23)$$

where

$$|F_{fi}|^2 = \frac{1}{2} \sum_r \sum_{r'} \left| u^{\dagger r'}(\mathbf{p}_f) u^r(\mathbf{p}_i) \right|^2$$

$$= \frac{1}{2} \sum_r \sum_{r'} \left| \bar{u}^{r'}(\mathbf{p}_f)\gamma^0 u^r(\mathbf{p}_i) \right|^2 \qquad (20.24)$$

where we have taken the average over initial spins and sum over final spins. Thus

$$|F_{fi}|^2 = \frac{1}{2} \sum_r \sum_{r'} [\bar{u}^{r'}(\mathbf{p}_f)\gamma^0 u^r(\mathbf{p}_i)][\bar{u}^r(\mathbf{p}_i)(\gamma^0) u^{r'}(\mathbf{p}_f)]$$

$$= \frac{1}{2} \sum_r \sum_{r'} [\bar{u}^{r'}_\alpha(\mathbf{p}_f)(\gamma^0)_{\alpha\beta} u^r_\beta(\mathbf{p}_i)\bar{u}^r_\lambda(\mathbf{p}_i)(\gamma^0)_{\lambda\rho} u^{r'}_\rho(\mathbf{p}_f)]$$

$$= \frac{1}{2} \sum_r \sum_{r'} u^{r'}_\rho(\mathbf{p}_f)\bar{u}^{r'}_\alpha(\mathbf{p}_f)(\gamma^0)_{\alpha\beta} u^r_\beta(\mathbf{p}_i)\bar{u}^r_\lambda(\mathbf{p}_i)(\gamma^0)_{\lambda\rho}$$

$$= \frac{1}{2}\left[\left(\frac{\not{p}_f + mc}{2mc}\right)_{\rho\alpha} (\gamma_0)_{\alpha\beta} \left(\frac{\not{p}_i + mc}{2mc}\right)_{\beta\lambda} (\gamma_0)_{\lambda\rho} \right]$$

$$= \frac{1}{8m^2c^2} Tr[(\not{p}_f + mc)\gamma^0(\not{p}_i + mc)\gamma^0]$$

$$= \frac{1}{8m^2c^2}[m^2c^2 Tr(\gamma^0\gamma^0) + Tr(\gamma^\mu\gamma^0\gamma^\nu\gamma^0)p_{f\mu}p_{i\nu}]$$

$$= \frac{4}{8m^2c^2}[m^2c^2 + (g^{\mu 0}g^{\nu 0} - g^{00}g^{\mu\nu} + g^{\nu 0}g^{\mu 0})p_{f\mu}p_{i\nu}]$$

$$= \frac{1}{2m^2c^2}[m^2c^2 + 2p_f^0 p_i^0 - p_f \cdot p_i]$$

$$= \frac{1}{2m^2c^2}\left[m^2c^2 + \frac{2E_f E_i}{c^2} - \frac{E_f E_i}{c^2} + \mathbf{P}_f \cdot \mathbf{P}_i\right]$$

$$= \frac{1}{2m^2c^2}\left[m^2c^2 + \frac{E^2}{c^2} + |\mathbf{p}|^2 \cos\theta\right]$$

$$= \frac{1}{2m^2c^4}[E^2 - c^2\mathbf{p}^2 + E^2 + c^2|\mathbf{p}|^2 \cos\theta]$$

$$= \frac{E^2}{m^2c^4}\left[1 - \frac{c^2\mathbf{p}^2}{E^2}\sin^2\frac{\theta}{2}\right]$$

$$= \frac{E^2}{m^2c^4}\left[1 - \beta^2 \sin^2\frac{\theta}{2}\right]$$

Hence finally from Eq. (20.23) and (20.25), we get

$$\frac{d\sigma}{d\Omega} = \frac{4Z^2e^4m^2}{|\mathbf{P}_i - \mathbf{P}_f|^4}\frac{E^2}{m^2c^4}\left(1 - \beta^2\sin^2\frac{\theta}{2}\right)$$

$$= \frac{4Z^2e^4}{(c\mathbf{q})^4}E^2\left(1 - \beta^2\sin^2\frac{\theta}{2}\right)$$

$$= \frac{(Z^2e^4)E^2}{4|c\mathbf{p}|^4\sin^4\frac{\theta}{2}}\left(1 - \beta^2\sin^2\frac{\theta}{2}\right)$$

$$= \frac{Z^2\alpha^2 E^2(\hbar E)^2 c^2}{4|c\mathbf{p}|^4\sin^4\frac{\theta}{2}}\left(1 - \beta^2\sin^2\frac{\theta}{2}\right)$$

In the high energy limit $\beta \to 1$, $c^2\mathbf{p}^2 \to E^2$

$$\frac{d\sigma}{d\Omega} = \frac{Z^2\alpha^2(\hbar c)^2}{E^2}\frac{\cos^2\frac{\theta}{2}}{4\sin^4\frac{\theta}{2}}$$

In the non-relativistic limit $\beta = v/c \to 0$, $c^2\mathbf{p}^2 \ll m^2c^4$, $E \approx m^2c^2$

$$\frac{d\sigma}{d\Omega} = \frac{Z^2e^4m^2}{4\mathbf{p}^4\sin^4\frac{\theta}{2}}$$

which is Rutherford scattering cross section.

Q3 Verify the identity

$$[A, BC] = \{A, B\}C - B\{A, C\}$$

Using the above identity, show that

$$[AB, CD] = A\{B, C\}D - AC\{B, D\}$$
$$+ \{A, C\}DB - C\{A, D\}B$$

Solution:

$$[AB, CD] = A[B, CD] + [A, CD]B$$

$$= A(\{B, C\}D - C\{B, D\})$$

$$+ (\{A, C\}D - C\{A, D\})B$$

$$= A\{B, C\}D - AC\{B, D\}$$

$$+ \{A, C\}DB - C\{A, D\}B$$

Q4 Out of the 16 Dirac matrices, the unit matrix commutes with all the 15 Dirac matrices. Show that 15 Dirac matrices generate the algebra of non compact SU(4) *viz.* Show that

$$[\gamma_\mu, \gamma_5] = 2\gamma_\mu\gamma_5$$

$$[\gamma_\mu\gamma_5, \gamma_5] = 2\gamma_\mu$$

$$[\gamma_\mu\gamma_5, \gamma_\nu\gamma_5] = 2i\sigma_{\mu\nu}$$

$$[\sigma_{\mu\nu}, \gamma_5] = 0$$

$$[\gamma_\mu, \sigma_{\lambda\rho}] = 2i(g_{\mu\lambda}\gamma_\rho - g_{\mu\rho}\gamma_\lambda)$$

$$[\gamma_\mu\gamma_5, \sigma_{\lambda\rho}] = 2i(g_{\mu\lambda}\gamma_\rho\gamma_5 - g_{\mu\rho}\gamma_\lambda\gamma_5)$$

$$[\sigma_{\mu\nu}, \sigma_{\lambda\rho}] = 2i(g_{\nu\lambda}\sigma_{\mu\rho} - g_{\mu\lambda}\sigma_{\nu\rho} + g_{\mu\rho}\sigma_{\nu\lambda} - g_{\nu\rho}\sigma_{\mu\lambda})$$

Solution: Using the solution of Q3 and the relations $\{\gamma_\mu, \gamma_\nu\} = 2g_{\mu\nu}, \gamma_\mu\gamma_5 = -\gamma_5\gamma_\mu$ it is trivial to derive the above commutation relation. For instance

$$[\gamma_\mu, \sigma_{\lambda\rho}] = \frac{i}{2}[\gamma_\mu, \gamma_\lambda\gamma_\rho - \gamma_\rho\gamma_\lambda]$$

$$= \frac{i}{2}[\gamma_\mu, \gamma_\lambda\gamma_\rho] - \frac{i}{2}[\gamma_\mu, \gamma_\rho\gamma_\lambda]$$

$$= \frac{i}{2}\{\gamma_\mu, \gamma_\lambda\}\gamma_\rho - \frac{i}{2}\gamma_\lambda\{\gamma_\mu, \gamma_\rho\} - \frac{i}{2}(\lambda \leftrightarrow \rho)$$

$$= \frac{2i}{2}(g_{\mu\lambda}\gamma_\rho - g_{\mu\rho}\gamma_\lambda) - \frac{i}{2}(\lambda \leftrightarrow \rho)$$

$$= 2i(g_{\mu\lambda}\gamma_\rho - g_{\mu\rho}\gamma_\lambda)$$

$$[\sigma_{\mu\nu}, \sigma_{\lambda\rho}] = \frac{i}{2}\left(\frac{i}{2}\right)[(\gamma_\mu\gamma_\nu - \gamma_\nu\gamma_\mu), (\gamma_\lambda\gamma_\rho - \gamma_\rho\gamma_\lambda)]$$

$$= \frac{-1}{4}[(2\gamma_\mu\gamma_\nu - 2g_{\mu\nu}), (2\gamma_\lambda\gamma_\rho - 2g_{\rho\mu})]$$

$$= -[\gamma_\mu\gamma_\nu, \gamma_\lambda\gamma_\rho]$$

$$= -(\gamma_\mu\{\gamma_\nu, \gamma_\lambda\}\gamma_\rho - \gamma_\mu\gamma_\lambda\{\gamma_\nu, \gamma_\rho\}$$
$$+ \{\gamma_\mu, \gamma_\lambda\}\gamma_\rho\gamma_\nu - \gamma_\lambda\{\gamma_\mu, \gamma_\rho\}\gamma_\nu)$$

$$= -(2g_{\nu\lambda}\gamma_\mu\gamma_\rho - \gamma_\mu\gamma_\lambda 2g_{\nu\rho}$$
$$+ 2g_{\mu\lambda}\gamma_\rho\gamma_\nu - 2g_{\mu\rho}\gamma_\lambda\gamma_\nu)$$

$$= -2(g_{\nu\lambda}(g_{\mu\rho} - i\sigma_{\mu\rho}) - g_{\nu\rho}(g_{\mu\lambda} - i\sigma_{\mu\lambda})$$
$$+ 2g_{\rho\nu} - i\sigma_{\rho\nu} - 2g_{\mu\rho}(g_{\lambda\mu} - i\sigma_{\lambda\nu}))$$

$$= 2i(g_{\nu\lambda}\sigma_{\mu\rho} - g_{\mu\lambda}\sigma_{\nu\rho} + g_{\mu\rho}\sigma_{\nu\lambda} - g_{\nu\rho}\sigma_{\mu\lambda})$$

Q5 Show that the Dirac Hamiltonian for hydrogen like atom in the non-relativistic approximation is given by

$$H = \frac{\widehat{\mathbf{p}}^2}{2m} + eA_o - \frac{\widehat{\mathbf{p}}^4}{8m^2c^2} - \frac{e\hbar\boldsymbol{\sigma}\cdot(\mathbf{E}\times\mathbf{p})}{4m^2c^2}$$

where $A_o = V(r) = -\frac{Ze}{r}$

Solution: We start from Eq. (20.182) of the text $(-\hbar^2\boldsymbol{\nabla}^2 = \widehat{\mathbf{p}}^2)$:

$$\left[2\frac{EA^o}{c^2} - \frac{e^2A_o^2}{c^2} + \widehat{\mathbf{p}}^2 \pm \frac{ie\hbar}{c}\frac{dA^o}{dr}\boldsymbol{\sigma}\cdot\mathbf{r}\right.$$
$$\left. - \left(\frac{E^2 - m^2c^4}{c^4}\right)\right]\phi_{L,R} = 0 \qquad (20.25)$$

Now

$$E = \epsilon + mc^2$$

and the electric field

$$\mathbf{E} = -\frac{1}{r}\frac{d\mathrm{A}^\circ}{dr}\mathbf{r}$$

and the above equation can be written as

$$\left[\frac{\widehat{\mathbf{p}}^2}{2m} - (\epsilon - eA_o) - \frac{1}{2mc^2}(\epsilon - eA_o)^2 \mp \frac{ie\hbar}{2mc}\boldsymbol{\sigma}\cdot\mathbf{E}\right]\phi_{L,R} = 0$$

$$(20.26)$$

To go to the non-relativistic limit, it is convenient to work with $\phi_{A,B}$ rather than $\phi_{L,R}$. Now in the Chiral representation of $\gamma-$ matrices,

$$\psi = \frac{1+\gamma^o}{2}\psi_A + \frac{1-\gamma^o}{2}\psi_B$$

$$= \frac{1}{2}\begin{pmatrix} 1 & 1 \\ 1 & 1 \end{pmatrix}\begin{pmatrix} \phi_A \\ 0 \end{pmatrix} + \frac{1}{2}\begin{pmatrix} 1 & -1 \\ -1 & 1 \end{pmatrix}\begin{pmatrix} 0 \\ \phi_B \end{pmatrix}$$

$$\phi_L = \frac{1}{2}(\phi_A - \phi_B), \quad \phi_R = \frac{1}{2}(\phi_A + \phi_B),$$

so that

$$\phi_A = \frac{1}{2}(\phi_L - \phi_R), \quad \phi_B = \frac{1}{2}(\phi_R + \phi_L),$$

Then from Eq. (20.26), by addition

$$\left[\frac{\widehat{\mathbf{p}}^2}{2m} - (\epsilon - eA_o) - \frac{1}{2mc^2}(\epsilon - eA_o)^2\right]\phi_A$$

$$+ \frac{ie\hbar}{2mc}\boldsymbol{\sigma}\cdot\mathbf{E}\phi_B = 0 \qquad (20.27)$$

Now from this equation and Eq. (20.269) of the text, we have

$$i\hbar\boldsymbol{\sigma}\cdot\boldsymbol{\nabla}\phi_A = -\frac{i\hbar}{mc}\frac{\partial}{\partial t}\phi_B - mc\phi_B$$

and in the presence of electrostatic field

$$\frac{\partial}{\partial t} \to \frac{\partial}{\partial t} + \frac{ie}{\hbar} A^o,$$

so that,

$$-\boldsymbol{\sigma} \cdot \widehat{\mathbf{p}} = \left[\left(-\frac{i\hbar}{c} \frac{\partial}{\partial t} + \frac{e}{c} A^o \right) + mc \right] \phi_B$$

For stationary states $\frac{\partial}{\partial t} \to -\frac{iE}{\hbar}$, so that

$$-\boldsymbol{\sigma} \cdot \widehat{\mathbf{p}} \phi_A = -\frac{1}{c} [E - eA_o + mc^2] \phi_B$$

Now $E = \epsilon + mc^2$, so that

$$\phi_B = \frac{1}{1 + \frac{1}{2mc^2}(\epsilon - eA_o)} \frac{\boldsymbol{\sigma} \cdot \widehat{\mathbf{p}}}{2mc} \phi_A \qquad (20.28)$$

We now assume $E \simeq mc^2$,

$$| eA_o | \ll mc^2, \quad \left(\frac{\epsilon - eA_o}{2mc^2} \right)$$

is roughly $\simeq \frac{v^2}{c^2}$, $E^i = -\frac{\partial A^o}{\partial x^i}$, i.e, $A^o \sim Ea$, a is the linear dimension of the system and is of the order of the Bohr radius for an atomic system. Thus

$$\frac{\hbar E}{mA_o} \sim \frac{\frac{\hbar}{a}}{m} \sim \frac{p}{m} \sim \frac{v}{c}, \quad \frac{e\hbar E}{m^2 c^3} \sim \frac{eA_o}{m^2 c^2} \frac{v}{c} \sim \frac{v^3}{c^3}$$

Substituting Eq. (20.28) in Eq. (20.27) and multiplying by $[1 - \frac{1}{2mc^2}(\epsilon - eA_o)]$, to the leading order

$$\left[\frac{\mathbf{p}^2}{2m} - (\epsilon - eA_o) - \frac{1}{4m^2 c^2}(\epsilon - eA_o)\mathbf{p}^2 \right.$$

$$\left. + \frac{ie\hbar}{2mc} \boldsymbol{\sigma} \cdot \mathbf{E} \frac{\boldsymbol{\sigma} \cdot \widehat{\mathbf{p}}}{2mc} \right] \phi_A = 0$$

Now

$$(\boldsymbol{\sigma} \cdot \mathbf{E})(\boldsymbol{\sigma} \cdot \mathbf{p}) = \sigma^i \sigma^j E^i \widehat{p}^j = (\delta^{ij} + \epsilon^{ijb} \sigma^b) E^i \widehat{p}_j$$

$$= \mathbf{E} \cdot \widehat{\mathbf{p}} + i\boldsymbol{\sigma} \cdot (\mathbf{E} \times \widehat{\mathbf{p}})$$

and the above equation becomes

$$\left[\frac{\mathbf{p}^2}{2m} + eA^o - \frac{(\epsilon - eA_o)}{4m^2c^2}\widehat{\mathbf{p}}^2 + ie\hbar\frac{\mathbf{E}\cdot\widehat{\mathbf{p}}}{4m^2c^2}\right.$$

$$\left. - \frac{e\hbar}{4m^2c^2}\boldsymbol{\sigma}\cdot(\boldsymbol{\sigma}\times\widehat{\mathbf{p}})\right]\phi_A = \epsilon\phi_A \qquad (20.29)$$

It cannot be regarded as an eigenvalue equation because it has three difficulties: (i) it contains a non-hermitian term, (ii) it contains ϵ on the left side, and (iii) there is difficulty with normalization, namely

$$\int(\phi_A^\dagger\phi_A + \phi_B^\dagger\phi_B)d^3x = 1$$

Now from Eq. (20.29), $\phi_B(\mathbf{x}) \simeq \frac{\boldsymbol{\sigma}\cdot\widehat{\mathbf{p}}}{2mc}\phi_A$ so that, using $\boldsymbol{\sigma}\cdot\widehat{\mathbf{p}}\boldsymbol{\sigma}\cdot\widehat{\mathbf{p}} = \widehat{\mathbf{p}}^2$, the L.H.S

$$= \int\left[\phi_A^\dagger(\mathbf{x})\phi_A(\mathbf{x}) + \phi_A^\dagger(\mathbf{x})\frac{\widehat{\mathbf{p}}^2}{4mc^2}\phi_A(\mathbf{x})\right]$$

This suggests in order to satisfy the normalization condition ϕ_A should be of the form

$$\phi_A = \left(1 - \frac{\widehat{\mathbf{p}}^2}{8m^2c^2}\right)\Phi,$$

so that to the order we are working

$$\int(\Phi^\dagger(\mathbf{x})\Phi(\mathbf{x}))d^3x = 1$$

Thus Eq. (20.30) becomes,

$$\left[\frac{\widehat{\mathbf{p}}^2}{2m} + eA^o - \frac{(\epsilon - eA_o)}{4m^2c^2}\widehat{\mathbf{p}}^2 - i\frac{\mathbf{E}\cdot\widehat{\mathbf{p}}}{4m^2c^2}\right.$$

$$\left. - \frac{e\hbar}{4m^2c^2}\boldsymbol{\sigma}\cdot(\mathbf{E}\times\widehat{\mathbf{p}})\right]\left(1 - \frac{\widehat{\mathbf{p}}^2}{8m^2c^2}\right)\Phi$$

$$= \epsilon\left(1 - \frac{\widehat{\mathbf{p}}^2}{8m^2c^2}\right)\Phi$$

Multiplying both sides by $(1 - \frac{\widehat{\mathbf{p}}^2}{8m^2c^2})$, to the order we are working

$$
\left[\frac{\widehat{\mathbf{p}}^2}{2m} - \frac{\widehat{\mathbf{p}}^4}{8m^2c^c} + eA^o - \frac{e}{8m^2c^2}(\widehat{\mathbf{p}}^2 A^o - A^o\widehat{\mathbf{p}}^2) \right.
$$

$$
\left. - ie\hbar \frac{\mathbf{E} \cdot \mathbf{p}}{4m^2c^2} - \frac{e\hbar}{4m^2c^2}\boldsymbol{\sigma} \cdot (\mathbf{E} \times \widehat{\mathbf{p}}) \right] \Phi = \epsilon\Phi
$$

$$(20.30)$$

Now

$$
[\widehat{\mathbf{p}}^2, A^o]f = \widehat{\mathbf{p}}[\widehat{\mathbf{p}}, A^o]f + [\widehat{\mathbf{p}}, A^o]\widehat{\mathbf{p}}f
$$
$$
= \widehat{\mathbf{p}}[-i\hbar\boldsymbol{\nabla}A^o]f + [-i\hbar\boldsymbol{\nabla} \cdot A^o]\widehat{\mathbf{p}}f
$$
$$
-\hbar^2\boldsymbol{\nabla}^2 A^o f + (-i\hbar\boldsymbol{\nabla}A^o) \cdot (-i\hbar\boldsymbol{\nabla})f + (-i\hbar\boldsymbol{\nabla}A^o) \cdot \widehat{\mathbf{p}}f
$$
$$
= [\hbar^2\boldsymbol{\nabla} \cdot \mathbf{E} - 2i\hbar\mathbf{E} \cdot \widehat{\mathbf{p}}]f
$$

Thus Eq. (20.31) becomes,

$$
\left[\frac{\widehat{\mathbf{p}}^2}{2m} - \frac{\widehat{\mathbf{p}}^4}{8m^2c^c} + eA^o - \frac{e\hbar^2}{8m^2c^2}\boldsymbol{\nabla} \cdot \mathbf{E} \right.
$$

$$
\left. - \frac{e\hbar}{4m^2c^2}\boldsymbol{\sigma} \cdot (\mathbf{E} \times \widehat{\mathbf{p}}) \right] \Phi = \epsilon\Phi \qquad (20.31)
$$

or

$$
H\Phi = \epsilon\Phi
$$
$$
H = \left[\frac{\widehat{\mathbf{p}}^2}{2m} + eA^o - \frac{\widehat{\mathbf{p}}^4}{8m^2c^c} - \frac{e\hbar^2}{8m^2c^2}\boldsymbol{\nabla} \cdot \mathbf{E} \right.
$$

$$
\left. - \frac{e\hbar}{4m^2c^2}\boldsymbol{\sigma} \cdot (\mathbf{E} \times \widehat{\mathbf{p}}) \right] \qquad (20.32)
$$

We now discuss the physical significance of each term. The first two terms appear in the non-relativistic equation i.e the Schrödinger equation. The third term is the relativistic correction to kinetic energy.

$$\sqrt{m^2c^4 + |\mathbf{p}|^2c^2} - mc^2 = \frac{|\mathbf{p}|^2}{2m} - \frac{|\mathbf{p}|^4}{8m^2c^2} + \cdots$$

The last term in Eq. (20.33) represents the spin interaction of the moving electron with the electric field and is called the Thomas term, who considered it before Dirac theory. For a central potential,

$$eA^o = V(r), \quad \mathbf{E} = -\frac{dV}{dr}\frac{\mathbf{r}}{r}$$

and we obtain

$$-\frac{e\hbar}{4m^2c^2}\boldsymbol{\sigma}\cdot(\mathbf{E}\times\widehat{\mathbf{p}}) = -\frac{\hbar}{4m^2c^2}\left(-\frac{1}{r}\frac{dV}{dr}\right)\boldsymbol{\sigma}\cdot(\mathbf{r}\times\widehat{\mathbf{p}})$$

$$= \frac{1}{2m^2c^2}\frac{1}{r}\frac{dV}{dr}\mathbf{S}\cdot\mathbf{L}$$

where $\mathbf{S} = \frac{\hbar}{2}$, and $\mathbf{L} = \mathbf{r}\times\mathbf{p}$. Thus the well known spin orbit force

$$H_{SL} = \frac{1}{2m^2c^2}\frac{1}{r}\frac{dV}{dr}\mathbf{S}\cdot\mathbf{L}$$

in atomic physics is an automatic consequence in Dirac theory. As far as the fourth term in Eq. (20.33) (known as Darwan term) is concerned, we note the it is just 4π times the charge density. For the hydrogen atom

$$\boldsymbol{\nabla}\cdot\mathbf{E} = -4\pi e\delta^3(\mathbf{r})$$

Q6 For the Dirac Hamiltonian given in the previous question, find the energy levels for hydrogen like atom for which

$$eA_o = V(r) = -\frac{e^2}{r}$$

Solution: We write the Dirac Hamiltonian as

$$H = H_o + H_1 + H_D$$

where

$$H_1 = -\frac{1}{8}\frac{\widehat{\mathbf{p}}^4}{m^3 c^2} + H_{SL}, \quad H_D = \frac{4\pi e^2 \hbar^2}{8m^2 c^2}\delta^3(\mathbf{r}) \quad (20.33)$$

We have H_1 and H_D as perturbations to

$$H_o = \frac{\widehat{\mathbf{p}}}{2m} - \frac{e^2}{r}$$

and the wave functions for the hydrogen atom in non-relativistic quantum mechanics are used as the unperturbed wave function.

Then for the hydrogen atom, H_D gives an energy shift

$$\int \frac{e\hbar^2}{8m^2 c^2} 4\pi e \delta^3(r) |\psi^{Sch}(r)|^2 d^3 r = 4\pi \frac{e^2 \hbar^2}{8m^2 c^2}|\psi_{nlm}(0)|^2$$

This is non-vanishing only for the s-state. For the s-state from Eq. (7.47) of the text

$$\psi_{n00}(0) = -\left(\left(\frac{2}{na_0}\right)^3 \frac{(n-1)!}{2n(n!)^3}\right)^{\frac{1}{2}} \frac{(n!)^2}{(n-1)!} Y_{00}(0,0)$$

Thus

$$|\psi_{n00}(0)|^2 = \frac{4}{n^3 a_0^3 \frac{1}{4\pi}}$$

Hence, the energy shift in the energy levels of hydrogen atom due to the Darwin term is given by

$$4\pi \frac{e^2 \hbar^2}{8m^2 c^2} \frac{4}{n^3 a_0^3} \frac{1}{4\pi} = -E_n^{(0)} \frac{\alpha^2}{n} \quad (20.34)$$

where $\alpha = \frac{e^2}{\hbar c}$ and

$$E_n^{(0)} = -\frac{e^2}{2a_0 n^2}, \quad a_0 = \frac{\hbar^2}{me^2}$$

Now we evaluate the contributions to the energy levels due to H_1.

Now $-\frac{\hat{\mathbf{p}}^4}{8m^3c^2}$ commutes with all the angular momentum operators. However, H_{SL} commutes with L^2, J^2, J_z (note that $\mathbf{J}^2 = \mathbf{L}^2 + \mathbf{S}^2 + 2\mathbf{S} \cdot \mathbf{L}$).

Thus we choose this set to label the basis vectors $|ljm>$.

Now

$$\mathbf{S} \cdot \mathbf{L} = \frac{1}{2}(\mathbf{J}^2 - \mathbf{L}^2 - \mathbf{S}^2) \qquad (20.35)$$

Therefore

$$\langle ljm|\mathbf{S} \cdot \mathbf{L}|ljm\rangle = \langle \mathbf{S} \cdot \mathbf{L}\rangle_{lj}$$

$$= \frac{\hbar^2}{2}\left[j(j+1) - l(l+1) - \frac{3}{4}\right] \qquad (20.36)$$

for the diagonal matrix elements (all other matrix elements vanish).

Now $j = l + \frac{1}{2}$ or $l - \frac{1}{2}$. Thus

$$\langle \mathbf{S} \cdot \mathbf{L}\rangle_{ij} = \frac{\hbar^2}{2}l, \quad j = l + \frac{1}{2}$$

$$= \frac{\hbar^2}{2}(l+1), \quad j = l - \frac{1}{2} \qquad (20.37)$$

for non-vanishing diagonal matrix elements. Also we note that

$$\langle \mathbf{S} \cdot \mathbf{L}\rangle = 0, \quad \text{if } l = 0$$

The zeroth order eigenstates for the hydrogen atom can be written as

$$|nljm\rangle = |nl\rangle|ljm\rangle$$

where $|nl\rangle$ refers to the radial motion:

$$\langle r|nl\rangle = R_{nl}(r)$$

The $R_{nl}(r)$'s are normalised radial eigenfunctions

$$\int_0^\infty R_{n'l'}(r)R_{nl}(r)r^2dr = \delta_{nn'}\delta_{ll'}$$

Now

$$V = -\frac{e^2}{r^3}$$

$$\frac{1}{r}\frac{dV}{dr} = \frac{e^2}{r^3}$$

Thus the only non-vanishing matrix elements of H_{SL} are

$$\langle nljm|H_{SL}|nljm\rangle$$

$$= \frac{e^2}{2m^2c^2}\langle nl|\frac{1}{r^3}|nl\rangle\langle ljm|\mathbf{S}\cdot\mathbf{L}|ljm\rangle\frac{e^2}{2m^2c^2}\langle nl|\frac{1}{r^3}|nl\rangle$$

$$\times \begin{pmatrix} \frac{\hbar^2}{2}l, & j = l + \frac{1}{2} \\ -\frac{\hbar^2}{2}(l+1), & j = 1 - \frac{1}{2} \end{pmatrix} \tag{20.38}$$

We note that $\widehat{\mathbf{p}}^4$ commutes with $\mathbf{J}^2, \mathbf{L}^2$ and $\mathbf{S}^2$. Thus the non-vanishing matrix elements of $-\frac{\widehat{\mathbf{p}}^4}{8m^3c^2}$ is given by

$$\left\langle -\frac{\widehat{\mathbf{p}}^4}{8m^3c^2} \right\rangle_{nljm}$$

$$= -\frac{1}{2mc^2}\langle nl|(H_0 + \frac{e^2}{r})^2|nl\rangle$$

$$= -\frac{1}{2mc^2}\langle nl| \left(E_n^{(0)} + \frac{e^2}{r} \right)^2 |nl\rangle$$

$$= -\frac{1}{2mc^2}\left\{ E_n^{(0)2} + 2E_n^{(0)}e^2\langle nl|\frac{1}{r}|nl\rangle + e^4\langle nl|\frac{1}{r^2}|nl\rangle \right\} \tag{20.39}$$

Now from the table of the radial integrals involving hydrogen wave functions [see for example, Atomic Spectra, Condon and Shortley, p. 117],

$$\left\langle nl \left| \frac{1}{r} \right| nl \right\rangle = \frac{1}{a_0 n^2} \tag{20.40a}$$

$$\left\langle nl \left| \frac{1}{r^2} \right| nl \right\rangle = \frac{1}{a_0^2 n^3}\frac{1}{(l + \frac{1}{2})} \tag{20.40b}$$

$$\left\langle nl \left| \frac{1}{r^3} \right| nl \right\rangle = \frac{1}{a_0^3 n^3}\frac{1}{(1 + l)(l + \frac{1}{2})l} \tag{20.40c}$$

Hence from Eq. (20.40) using Eqs. (20.41), we have

$$
\left\langle -\frac{\widehat{\mathbf{p}}^4}{8m^3c^2} \right\rangle_{nljm} = \frac{1}{2mc^2} E_n^{(0)} \frac{2e^2}{a_0 n^2} \left(\frac{1}{4} - 1 + \frac{n}{l+\frac{1}{2}} \right)
$$

$$
= E_n^{(0)} \left(\frac{e^2}{\hbar c} \right)^2 \frac{1}{n^2} \left(\frac{n}{l+\frac{1}{2}} - \frac{3}{4} \right)
$$

$$\tag{20.41}$$

Similarly, using Eq. (20.41c), we have from Eq. (20.39),

$$
\langle nljm|H_{SL}|nljm \rangle = -E_n^{(0)} \left(\frac{e^2}{\hbar c} \right)^2 \frac{1}{n} \frac{1}{(2l+1)}
$$

$$
\times \left(\begin{array}{cc} \frac{1}{l+1}, & j = l + \frac{1}{2} \\ -\frac{1}{l}, & j = l - \frac{1}{2} \end{array} \right)
$$

$$\tag{20.42}$$

for $l \neq 0$. For $l = 0$ we have the Darwin term (20.35).
Now putting $\frac{e^2}{\hbar c} = \alpha$ and combining Eqs. (20.30), (20.42) and (20.43), we see that a shift in the energy levels due to perturbation H_1 is given by

$$
\triangle E_n^{(0)} = \langle nljm|H_1|nljm \rangle
$$

$$
= E_n^{(0)} \frac{\alpha^2}{n^2} \left(-\frac{3}{4} + \frac{n}{2l+1} \right)
$$

$$
+ \frac{1}{2l+1} \left(\begin{array}{cc} \left(-\frac{n}{l+1}\right), & j = l + \frac{1}{2} \\ \left(\frac{n}{l}\right), & j = l - \frac{1}{2} \end{array} \right)
$$

$$
= E_n^{(0)} \frac{\alpha^2}{n^2} \left(\frac{n}{j+\frac{1}{2}} - \frac{3}{4} \right)
$$

$$\tag{20.43}$$

Hence the energy levels for the hydrogen atom are given by

$$
E_n = E_n^{(0)} + \triangle E_n^{(0)}
$$

$$
= E_n^{(0)} \left(1 + \frac{\alpha^2}{n} \left(\frac{1}{j+\frac{1}{2}} - \frac{3}{4n} \right) \right)
$$

$$\tag{20.44}$$

This agrees with Eq. (20.200) in the text, obtained from the exact treatment for $\alpha \ll 1$.

Chapter 21

Dirac Equation in $(1+2)$ Dimensions: Application to Graphene

Q21.1 Show that

(i)

$$\sigma^{\mu\nu} = \epsilon^{\lambda\mu\nu}\gamma_\lambda$$

$$\mu, \nu = 0, 1, 2. \quad \epsilon^{012} = 1, \quad \sigma^{\mu\nu} = \frac{1}{2i}(\gamma^\mu\gamma^\nu - \gamma^\nu\gamma^\mu)$$

(ii)

$$Tr[\gamma^\mu\gamma^\nu\gamma^\lambda] = -2i\epsilon^{\mu\nu\lambda}$$

Solution: We choose the following irreducible representation of γ-matrices

$$\gamma^0 = \sigma^3, \quad \gamma^1 = i\sigma^1, \quad \gamma^2 = i\sigma^2$$

Using the properties of Pauli matrices,

$$\gamma^{0^2} = 1, \quad \gamma^{1^2} = -1, \quad \gamma^{2^2} = -1$$

$$\gamma^1\gamma^2 = -\sigma^1\sigma^2 = -i\sigma^3 = -i\gamma^0$$

$$\gamma^0\gamma^1 = -\sigma^2 = i\gamma^2 = -i\gamma_2$$

$$\gamma^0\gamma^2 = -i\gamma^1$$

All the above relations can be written in the compact form

$$\gamma^\mu\gamma^\nu = g^{\mu\nu} - i\epsilon^{\mu\nu\lambda}\gamma_\lambda$$

Then using this relation

$$Tr(\gamma^\mu\gamma^\nu\gamma^\lambda) = Tr\left[(g^{\mu\nu}\gamma_\lambda - i\epsilon^{\mu\nu\alpha}\gamma_\alpha)\gamma^\lambda\right]$$
$$= -i\epsilon^{\mu\nu\alpha}Tr(\gamma_\alpha\gamma^\lambda), \quad Tr(\gamma^\lambda) = 0$$
$$= -i\epsilon^{\mu\nu\alpha}(2\delta_\alpha^\lambda)$$
$$= -2i\epsilon^{\mu\nu\lambda}$$

Q21.2 For the static magnetic field:

$$A^0 = 0, \quad \overset{i}{\dot{A}} = 0, \quad \mathbf{E} = 0$$

show that Eq. (21.60) of the text takes the form for $\sigma^3 = 1$

$$\left[-D_1^2 - D_2^2 - \frac{e}{c}B\right]\phi_+ = \frac{\epsilon^2}{v_f^2}\phi_+$$

Show that

$$[D_1, D_2] = -\frac{ie}{c}B$$

Define Canonical variables $\hat{Q}, \hat{P}$:

$$-i\frac{D_2}{\frac{eB}{c}} = \hat{Q}, \quad iD_1 = \hat{P},$$

so that

$$[\hat{Q}, \hat{P}] = i$$

and

$$\left[\frac{1}{2}\hat{P}^2 + \frac{1}{2}\omega^2\hat{Q}^2\right]\phi_+ = E\phi_+$$

where

$$E = \frac{1}{2}\left(\frac{\epsilon^2}{v_f^2} + \frac{eB}{c}\right), \quad \omega = \frac{eB}{c}$$

The above is the equation of one-dimensional harmonic oscillator, and give energy levels

$$E = \left(n + \frac{1}{2}\right)\omega$$

$$\epsilon_\pm(n) = \pm\, \omega_c \sqrt{|n|}, \quad n = 0, \pm1, \pm2\cdots$$

$$\omega_c = \sqrt{2} v_f \sqrt{\frac{e}{c}B}$$

Solution: From Eq. (21.68) of the text and for the case given in the problem,

$$\left[-D_1^2 - D_2^2 - \frac{e}{c}B \right]\phi_+ = \frac{\epsilon^2}{v_f^2}\phi_+ \tag{21.1}$$

Now

$$
\begin{aligned}
[D_1, D_2] &= \frac{ie}{c}\left[\frac{\partial A_2}{\partial x^1} - \frac{\partial A_1}{\partial x^2} \right] \\
&= -\frac{ie}{c}\left[\frac{\partial A^2}{\partial x^1} - \frac{\partial A^1}{\partial x^2} \right] \\
&= -\frac{ie}{c}B
\end{aligned}
$$

Then

$$[\hat{Q}, \hat{P}] = \frac{c}{eB}[D_2, D_1] = i$$

Thus Eq. (21.1) can be written as

$$\left[\frac{1}{2}\hat{P}^2 + \frac{1}{2}\omega^2\hat{Q}^2 \right]\phi_+ = E\phi_+$$

where

$$E = \frac{1}{2}\left(\frac{\epsilon^2}{v_f^2} + \frac{eB}{c} \right), \quad \omega = \frac{eB}{c}$$

This is the equation of the one-dimensional harmonic oscillator, and gives energy levels

$$E = \left(n + \frac{1}{2} \right)\omega$$

or

$$\epsilon_\pm(n) = \pm\omega_c\sqrt{|n|}, \quad n = 0, \pm1, \pm2, \cdots$$

$$\omega_c = \sqrt{2}v_f \sqrt{\frac{eB}{c}}$$

Erratum for Quantum Mechanics (2nd Edition) by Fayyazuddin and Riazuddin

Chapter 1

- On page 13, in the line after Eq. (1.42), "Bohr atom" should be "Bohr radius".

Chapter 2

- On page 23, in the line just before Eq. (2.23), Eq. (2.20) should be Eq. (2.22).
- On page 24, in the line just after Eq. (2.28), Eq. (2.30) should be Eq. (2.27).
- On page 24, in the line just after Eq. (2.30), Eq. (2.31) should be Eq. (2.28).
- On page 25, in the lines just after Eq. (2.34), Schrodinger should be Schrödinger.
- On page 27, in the line just after Eq. (2.56), Eq. (2.53) should be Eq. (2.54).
- On page 27, in the paragraph just before Eq. (2.60), Eq. (2.58) should be Eq. (2.59).
- On page 29, the second equation after Eq. (2.65) should be
$$(\frac{\partial}{\partial x}x)\psi(x) = \psi(x) + x\frac{\partial}{\partial x}$$

Chapter 3

- On page 35 replace all bras "$\langle$" should be "(" and all kets "$\rangle$" should be ")".
- On page 40 in the second line Eq. (3.38) should be Eq. (3.35).
- On page 41 in the heading of part (c) there should be a colon : after δ-**function**
- On page 50, the last equation in Problem 3.8,

$$\frac{\sin^4\left(\left(\frac{p}{\hbar}-n\right)\frac{\pi}{2}\right)}{\left(\frac{\frac{p}{\hbar}-n}{2}\right)^2} \rightarrow \frac{\sin^2\left(\left(\frac{p}{\hbar}-n\right)\frac{\pi}{2}\right)}{\left(\frac{p^2}{\hbar^2}-n^2\right)^2}$$

Chapter 5

- On page 84, Problem 5.6 the last term of the equation

$$\left(\frac{2\beta^3}{\sqrt{\pi}}\right)^{\frac{1}{2}} e^{\frac{-i3\omega t}{2}} \rightarrow \left(\frac{2\beta^3}{\sqrt{\pi}}\right)^{\frac{1}{2}} xe^{\frac{-i3\omega t}{2}}$$

Chapter 6

- On page 91, in the fifth equation from the top regarding $\hat{\phi}_1$, "θ" should be "ϕ".
- On page 97 in Eq. (6.42),

$$Y_{lm}^*(P,\phi) \rightarrow Y_{lm}^*(\theta,\phi)$$

Chapter 7

- On page 111 in the last term of Eq. (7.40),

$$(q-p)_q^p \rightarrow (q-p)L_q^p(y)$$

- On page 114 in the last line of Problem 7.1,

$$1027A° \rightarrow 1215A°$$

Chapter 8

- On page 117 in the last line of first paragraph,

$$\psi(\mathbf{r}_1,\mathbf{r}_1,t) \rightarrow \psi(\mathbf{r}_1,\mathbf{r}_2,t)$$

- On page 123 in the line before Eq. (8.20c), "Fig. 8.1" should be "Fig. 8.2".
- On page 124, "From Fig. 8.1a" should be "From Fig. 8.2a".
- On page 125 in the paragraph after Eq. (8.27), "(Fig. (8.2))." should be "(Fig. (8.3))."
- On page 127 in the last line of the second paragraph of Section 8.5, "(Fig. 8.3)" should be "(Fig. 8.1)".
- On page 142 in the second line of the paragraph after Eq. (8.119),

$$\frac{2}{\pi}\Gamma \rightarrow \frac{2}{\pi}\frac{1}{\Gamma} \quad \text{and} \quad \frac{1}{\pi}\Gamma \rightarrow \frac{1}{\pi}\frac{1}{\Gamma}$$

- On page 143 on y-axis of Fig. 8.6,

$$\frac{2}{\pi}\Gamma \rightarrow \frac{2}{\pi}\frac{1}{\Gamma} \quad \text{and} \quad \frac{1}{\pi}\Gamma \rightarrow \frac{1}{\pi}\frac{1}{\Gamma}$$

- On page 144 in Eq. (8.135),

$$x > 0 \rightarrow x > a$$

- On pages 145 and 146 in Eqs. (8.139)–(8.150),

$$\sin 2ka \rightarrow \sin 2Ka \qquad \cos 2ka \rightarrow \cos 2Ka$$
$$\tan ka \rightarrow \tan Ka \qquad \cot ka \rightarrow \cot Ka$$
$$\sin^2 2ka \rightarrow \sin^2 2Ka \qquad \cos^2 2ka \rightarrow \cos^2 2Ka$$

- On page 146 in the first line of Eq. (8.154),

$$\sin 2K \rightarrow \sin 2Ka$$

- On page 150 in the second line of first paragraph,

$$t_i(- \rightarrow \infty) \rightarrow t_i(\rightarrow -\infty)$$

- On page 150, Eq. (8.185a) should be

$$i\hbar \dot{a}_{\boldsymbol{pp}_i}(t) = \sum_{\boldsymbol{p}} a'_{\boldsymbol{pp}_i}(t) e^{\frac{-i}{\hbar}(E_p - E_{\acute{p}})t} \langle \acute{\boldsymbol{p}}|V| \rightarrow \boldsymbol{p}\rangle$$

$$\rightarrow$$

$$i\hbar \dot{a}_{\boldsymbol{p'p}_i}(t) = \sum_{\boldsymbol{p}} a_{\boldsymbol{pp}_i}(t) e^{\frac{-i}{\hbar}(E_p - E_{\acute{p}})t} \langle \acute{\boldsymbol{p}}|V|\boldsymbol{p}\rangle$$

- In the last equation of page 150,

$$a'_{\boldsymbol{pp}} \to a_{\boldsymbol{p'p}}$$

- On page 151 in Eq. (8.186),

$$a'_{\boldsymbol{pp}_i} \to a_{\boldsymbol{p'p}_i}$$

- On page 151 in the first line of Eq. (8.188),

$$a_{\boldsymbol{p}_f\boldsymbol{p}_i}(t) \to a_{\boldsymbol{p'}_f\boldsymbol{p}_i} \quad \text{and} \quad \int_\infty^\infty \to \int_{-\infty}^\infty$$

- On page 152 in the lines before Eq. (8.195), "where as" should be "whereas".

Chapter 9

- On page 178 in Eq. (9.25),

$$\sum_n \sum_m |a_n\rangle\langle a_m| \to \sum_n \sum_m |a_n\rangle \alpha_{mn} \langle a_m|$$

- On page 193 in the first line of Eq. (9.122),

$$\{Q^\dagger, Q^\dagger\} = 1 \to \{Q^\dagger, Q^\dagger\} = 0$$

- On page 194 in Problem 9.5 in Eq. (9.123),

$$\sqrt{\hbar/m\omega} \to i\sqrt{\hbar/m\omega}$$

- On page 195 in Problem 9.8 in Eq. (9.124),

$$\langle |\hat{x}^2|0\rangle \to \langle n|\hat{x}^2|0\rangle$$

Chapter 10

- On page 204, the first line of Eq. (10.29),

$$\frac{dF}{dt} = \sum_i \left(\frac{\partial f}{\partial q_i} \frac{\partial H}{\partial p_i} - \frac{\partial f}{\partial p_i} \frac{\partial f}{\partial q_i} \right)$$

$$\to$$

$$\frac{dF}{dt} = \sum_i \left(\frac{\partial F}{\partial q_i} \frac{\partial H}{\partial p_i} - \frac{\partial F}{\partial p_i} \frac{\partial H}{\partial q_i} \right)$$

- On page 204 in the first line of the paragraph after Eq. (10.30), Eq. (20.32) should be Eq. (10.29).
- On page 205 in the line after Eq. (10.32), Eq. (20.33) should be Eq. (10.32).
- On page 212 in the lines before Eq. (10.67), "Prob. 10.6" should be "Prob. 10.4".
- On page 220 in both equations after Eq. (10.125),

$$|\psi\rangle \to |\psi\rangle_s$$

- On page 223,

$$\left(\frac{2\pi i j \delta t \hbar}{(j+1)}\right)^{\frac{1}{2}} \to \left(\frac{2\pi i j \delta t \hbar}{(j+1)m}\right)^{\frac{1}{2}}$$

- On page 227 in the line after Eq. (10.151), Eq. (10.7.4) should be Eq. (10.149).
- On page 227 in Eq. (10.155),

$$\int_0^t \to \int_0^T$$

- On page 233 in Problem 10.7,

$$\frac{\iota\hbar}{m\omega \sin \omega t} \to \frac{i\hbar}{m\omega \sin \omega T}$$

Chapter 11

- On page 273 in the line after Eq. (11.167), "(see Sec. 17.6)" should be "(see Sec. 17.7)".
- On page 278 in the last equations of Problem 11.15,

$$\sigma \cdot p|\mathbf{p} \uparrow\rangle \to \sigma \cdot \mathbf{p}|\mathbf{p} \uparrow\rangle$$

- On page 279 in the last line, "single state" should be "singlet state".

Chapter 15

- On page 324 in Eq. (15.5),

$$(H_{11} + H_{22}) \to (H_{12} + H_{21})$$

- On page 333 in Eq. (15.53),

$$(m^*)_{\alpha\alpha'} \to (m^*)_{\alpha'\alpha} \quad \text{and} \quad \Gamma^*_{\alpha\alpha'} \to \Gamma^*_{\alpha'\alpha}$$

Chapter 17

- On page 361 in Eq. (17.24), $\vec{D}$ should be **D**.
- On page 365 in equation before Eq. (17.39). $\frac{\partial}{\partial Z}$ should be $\frac{\partial}{\partial z}$.
- On page 374 in Eq. (17.99a), E^{λ^*} should be ϵ^{λ^*}.
- On page 382 in the first line, Eq. (17.121e) should be Eq. (17.121).
- On page 385 in text $\sum^0$ should be Σ^0 and in the equations

$$\overset{0}{\sum} \to \Sigma^0$$

- On page 393 in Eq. (17.203), $\epsilon \to \epsilon$.
- On page 393 in the line above Eq. (7.206), Eq. (17.190b) should be Eq. (17.189b).
- On page 395 in Eq. (17.215) of Problem 17.2,

$$\hat{P} = \frac{1}{m\omega}\left(\hat{P}_y - \frac{e}{c}A_y\right)$$

$$\to$$

$$\hat{P} = \left(\hat{p}_y - \frac{e}{c}A_y\right)$$

and in Eq. (17.216),

$$[\hat{Q}, \hat{p}] = i\hbar \to [\hat{Q}, \hat{P}] = i\hbar$$

and in the equation before Eq. (17.220), $(\hat{p}_y - \frac{e}{c}) \to (\hat{p}_y + \frac{e}{c})$.
- On page 395 in Problem 17.3 in the last line of equation, the last two terms

$$\frac{e^2}{mc^2}\mathbf{A}^2 + 2e\phi \to \frac{e^2}{mc^2}\psi^*\psi\mathbf{A}^2 + 2e\psi^*\psi\phi$$

- On page 396 in the third equation from the top,

$$(\psi^*\nabla\psi - \psi\nabla\psi^*) \to (\psi^*\mathbf{A}\cdot\nabla\psi - \psi\mathbf{A}\cdot\nabla\psi^*)$$

- On page 396 in the second last equation of Problem 17.3,

$$= \frac{e}{2mc^2}\psi^*\psi\mathbf{A}^2 \to -\frac{e}{2mc^2}\psi^*\psi\mathbf{A}$$

- On page 399 equation in Problem 17.7, E should be ϵ.

Chapter 18

- On pages 430 and 431 in all Problems, $16 \to 18$.

Chapter 19

- On page 444 in Eq. (19.93) delete $\hat{C}^t \hat{B}^t$.

Chapter 20

- On page 451 in the second line of the first paragraph above Eq. (20.9), $L^\dagger \to L^\uparrow$ and in the third line, $L_+^\dagger \to L_+^\uparrow$.
- On page 458 in Eqs. (20.46) and (20.47), $\hat{p} \to \hat{\mathbf{p}}$.
- On page 459 in the first line, Eq. (20.51) should be Eq. (20.52).
- On page 459 in Eq. (20.54), $o \to 0$ in the first row, fourth column of β matrix.
- On page 460 in Eq. (20.60) in the definition of γ_5 put $-$ sign before i.
- On page 460 in the first and second lines of Eq. (20.62), γ^5 should be γ_5.
- On page 474 in Eq. (20.116a), first row, second column matrix element "0" should be "1" and first row third column matrix element "1" should be "0" of κ^1 matrix.
- On page 475 in the third line of Eq. (20.117), κ_k should be S_k.
- Everywhere in Section 20.15, "$/p$" should be "$\not{p}$".
- On page 496 in equation after Eq. (20.210), $u^\dagger \mathbf{p} \to u^\dagger(\mathbf{p})$.
- On page 504 in the first term of the last equation, $\phi_B \to -\phi_B$.
- On page 505 in the second equation, $+mc\phi_B$ should be $-mc\phi_B$.
- On page 505 in the second line of Eq. (20.272) and Eq. (20.275), phi_B should be ϕ_B.
- On page 507 in the last line of last paragraph, $(1 - \gamma_5) \to (1 + \gamma_5)$.
- On page 508 in part (i) of Problem 20.3, $\gamma^n u \to \gamma^\nu$, in part (ii), $\frac{1}{3!}$ should be $\frac{i}{3!}$ and in part (iv) delete i from right side of the equation.
- On page 510 in the equation of Problem 20.7, $V\psi \to \frac{1}{c}V\psi$.
- On page 510 in the equation of Problem 20.8, $\frac{2}{m} \to \frac{2}{mc}$.
- On page 510 in the last equation, $p^\lambda \to p_\nu$.

- On page 511 in the first equation of Problem 20.10

$$\frac{1}{2m}\left(\mathbf{p} - \frac{e}{c}\mathbf{A}\right)^2 - \frac{e\hbar}{2mc}\boldsymbol{\sigma}\cdot\mathbf{B} + eA_0\psi_A$$

$$\rightarrow \left[\frac{1}{2m}(\mathbf{p} - \frac{e}{c}\mathbf{A})^2 - \frac{e\hbar}{2mc}\boldsymbol{\sigma}\cdot\mathbf{B} + eA_0\psi_A\right]$$

- On page 511 in the second last line of Problem 20.10, $\psi_A^\dagger \sigma_k \psi_A \rightarrow$ $\psi_A^\dagger \sigma^k \psi_A$.
- On page 512 in Problem 20.12, the last equation should be

$$S = \frac{1}{\sqrt{2}}\begin{pmatrix} 1 & -1 \\ 1 & 1 \end{pmatrix}$$

Printed in the United States
By Bookmasters